SpringerBriefs in Psychology

SpringerBriefs in Theoretical Advances in Psychology

Series Editors

Jaan Valsiner, Aalborg University
Aalborg, Denmark

Carlos Cornejo, Escuela de Psicologia
Pontificia Universidad Católica de Chile
Santiago, Macul, Chile

SpringerBriefs in Theoretical Advances in Psychology will be an extension from the currently renovated Annals of Theoretical Psychology in the direction of bringing short, single (or multiple) authored theoretical advancements across all areas of psychology to the international audience. The focus is on the development of innovative theoretical approaches and their discussion. The Series will have a clearly defined international and interdisciplinary focus – even if it remains within the discipline of psychology.

Featuring compact volumes of 100 to 115 pages, each Brief in the Series is meant to provide a clear, visible, and multi-sided recognition of the theoretical efforts of scholars around the world. It is targeted to researchers, graduate students, and professionals in Post-BA level psychology, education, anthropology, and sociology.

Briefs are published as part of the Springer's eBook collection, with millions of users worldwide. In addition, Briefs are available for individual print and electronic purchase.

Vanessa Lux

The Neuron in Context

 Springer

Vanessa Lux
Faculty of Psychology
Ruhr University Bochum
Bochum, Germany

ISSN 2192-8363   ISSN 2192-8371   (electronic)
SpringerBriefs in Psychology
ISSN 2511-395X   ISSN 2511-3968
SpringerBriefs in Theoretical Advances in Psychology
ISBN 978-3-031-55231-1   ISBN 978-3-031-55229-8   (eBook)
https://doi.org/10.1007/978-3-031-55229-8

This Springer imprint is published by the registered company Springer Nature Switzerland AG
The registered company address is: Gewerbestrasse 11, 6330 Cham, Switzerland

If disposing of this product, please recycle the paper.

Series Preface

The New Look at the Neuron: Self-constraining of Flexibility Through Context

The main challenge for biological sciences is to understand how a system is relating to its context in ways that guarantee—simultaneously—its own stability and change. This is the central issue for all developing systems—be these genetic, neuronal, behavioral, cognitive, social, or, last but not least, political. Understanding such complexity is a major task for abstractive generalization in developmental sciences. The present book outlines the theoretical efforts towards such understanding in the core of the neurosciences, looking in a new way at the unity of the central part of our nervous system—the neuron—in its hyper-complex dependency upon the context of neuronal networks that operate rapidly across the hierarchy of the nervous system as a whole. This is made possible by the genetic basis—the developmental stability and transformation afforded by epigenetic processes at the level of the work of the DNA-RNA-Protein synthesis cycles. The unity of all developmental systems is in their generative innovation based on stability.

The central unit of our nervous system—the neuron—stands out as a fitting challenge for our theoretical efforts to understand unity in enormous diversity. The neuron is a crucial connecting unit of all of the neuronal system—yet it is a singular player in the hyper-large crowd of similar neurons situated all over the nervous system and closely dependent upon their contexts. Understanding the role of a single neuron in this hyper-crowd that makes up the structure of the nervous system is an analytic challenge. This book outlines, step by step, how that challenge can be approached. The result is a new look at the neuron—elaborated in Chap. 6 (Box 6.1.)—that concisely sets up the theoretical frame for neuroscience. Thinking through the synopsis given in Chap. 6, the crucial features of the new theory of the neuron appear.

First of all, there is the basis for flexibility. Each neuron's metabolism needs to be stabilized and protected from the constant information influx coming from its immediate environment. This cannot be accomplished through our old—causal—models of thought, but requires a turn to catalytic models (Cabell & Valsiner, 2014; Valsiner, 2019). It is through epigenetic mechanisms—rather than any causal action by genes—that this stabilization can be accomplished. The science of genetics needed to reach its own flexible re-construction into epigenetics to make understanding the functioning of the neuron in the crowd of its compatriots possible. The rapid connectivity within that crowd allows for pre-adaptive multi-functional capacity—the function of a neuron is in its propensity to very high flexibility in taking on various functions. The nervous system is a prime example of an open system where equipotentiality and equifinality reign (Driesch, 1901). In its pre-adaptive functions, we are probably better thinking in terms of *generative multipotentiality* of the neuron in its rapid feed-forward binding with its current context to enable innovation in the neuronal system. Neurons need to have plasticity bigger than that of other cells to be in a position to carry out these forward-focused adaptive functions. That is granted by the relevance of context—the neuron can function only in relation with the context. This again is a reminder for us that our understanding of the functioning needs to begin from the assumption of the open systems that depend on exchange relationships with their environments. Our contemporary neurosciences are an arena for basic theoretical ideas of natural philosophy now becoming meaningful in the empirical practices of science.

Vienna, Austria   Jaan Valsiner
December 2023

References

Cabell, K. R., & Valsiner, J. (Eds.). (2014). *The catalyzing mind: Beyond models of causality* (*Advances of Theoretical Psychology*, Vol. 11). Springer.
Driesch, H. (1901). *Die Organischen Regulationen*. Wilhelm Engelmann.
Valsiner, J. (2019). From causality to catalysis in the social sciences. In J. Valsiner (Ed.). *Social philosophy of science for the social sciences* (pp. 125–146). Springer.

Preface

As a discipline, psychology often looks anxiously at neuroscience and its findings and how they align with psychological theory. As basic elements of neural function, neurons have been considered the primary source of mental states and consciousness. Research in this perspective focused on pinpointing specific emotions and cognitive functions in individual neurons or small neural networks, describing memory and learning in terms of synaptic changes, and identifying neurotransmitter levels related to behavioral differences. Psychology was originally positioned to provide a different perspective on our mental states, one that was not captured by our neurophysiological knowledge of brain cells. If we could explain all psychological functions, including our subjective experiences from single neuron cell parameters, we could simply replace psychology with neuroscience.

Neurons have been a major focus of research in the past century, and a neural description of the mind seemed a real possibility for some time. Ever since neurons were first described, a wealth of knowledge has accumulated. Detailed information about their individual components, chemical and electrical properties, proteome, genome, and epigenome is now available. Through observations conducted in both in vitro and in vivo settings, we learned how neurons form intricate networks, interact with each other, and demonstrate significant levels of adaptability and plasticity, both structurally and functionally. However, the increasing knowledge about the interconnected nature of neural functions challenged the belief that individual neurons are the fundamental units and sources of neural and psychological functions. In light of these findings, neuroscience has largely abandoned the localizationist and mechanistic frameworks of the twentieth century. The plastic, embodied, and network characteristics of our nervous system are widely acknowledged, and systems theory approaches to consciousness dominate this field.

Nevertheless, the underlying neuron theory has not changed despite the shift to the network level. Originally outlined as "neuron doctrine," the notion that neurons are the unidirectional and atomic source of neural function still serves as the blueprint for how we conceptualize these excitable cells. While the neuron doctrine has laid the foundation for understanding neural function, its reductionist approach fails to account for the complex, interconnected nature of neural networks. Having accumulated a tremendous amount of knowledge about the neuron and its workings in the brain throughout the last century, neuroscience is now confronted with the limitations imposed by the neuron doctrine.

In the upcoming chapters, I examine these limitations of the neuron doctrine and its reductionist approach from different angles, considering the new knowledge on the brain's connectivity, plasticity, and systemic and embodied nature, and explore how neuroepigenetic information contributes to a new understanding of the neuron in this context. The first chapter dives into the growing theoretical divide between systemic and network approaches in neuroscience, and the underlying neuron theory. It explores the historical context of the neuron doctrine, the evolving network perspective, and various attempts to characterize and classify neurons. In this context, neuroepigenetic information provides further insights into the complexity of observed neuronal diversity by shedding light on the dynamic nature and need for functional stabilization of a single neuronal cell. The question arises as to whether the environmental sensitivity of epigenetic mechanisms indicates neuronal plasticity, or whether they serve as stabilizing mechanisms within the individual developmental pathways of a neuron. To answer this question, the second chapter addresses how plasticity is understood at different levels of brain function, from phenotypic plasticity to cognitive plasticity and neural plasticity, including synaptic and non-synaptic forms of plasticity. The role of epigenetic mechanisms in neural plasticity and their contribution to neuronal survival and adaptability in response to the ever-changing demands of a neuron's functional context are discussed.

The notion of plasticity as flexible adaptability at the single-cell and epigenetic level stands in contrast to concepts that have historically characterized psychobiological development, particularly the maturation of neural structures and functions as a sequential decrease in plasticity. In the third chapter, I discuss how both views can be reconciled by proposing canalization as a developmental mechanism that allows for the stabilization of function, while simultaneously preserving plasticity. Neuroepigenetic mechanisms show strong potential to function as biomarkers of canalization and plasticity in the brain. How stability and plasticity are achieved at the level of individual neurons requires further exploration. However, the concept of canalization shifts the focus to the cellular and wider context of the neuron. The fourth chapter examines the various contexts that impact a neuron's function, ranging from its immediate tissue surroundings to larger neural networks and the bodily and socio-cultural contexts of psychobiological development. A neuron is not an isolated entity but relies heavily on its surrounding cellular environment for survival and proper functioning. It is the result of the collective effects of a group of cells and neural networks. In addition, psychological processes and cultural conditions provide an even broader but similarly significant context within which individual neurons are recruited and carry out specific functions. In this multilayered context, a single neuron is exposed to a constant stream of information that must be integrated and decoded to perform its functions. Neuroepigenetic mechanisms enable the neuron to integrate these different signaling contexts in an energy-efficient manner and thus provide measurements to register neuron activity in context. The fifth chapter further outlines the different layers of information present in neuroepigenetic data and discusses how it registers the metabolic and activation history of a single neuron in an aggregated manner over time. The conditions and types of information for single-cell and whole-tissue epigenetic data, the temporal character of epigenetic data, and the use of genetic information to characterize neuroepigenetic information are discussed, and the application of information theory approaches to characterize the information structure of neuroepigenetic data is evaluated. The chapter concludes that functional molecular epigenetic measurements relevant to characterizing neural activity are more likely to be detected at the tissue level than at the single-cell level.

Finally, in the sixth chapter, I propose basic elements of a new neuron theory. This new neuron theory depicts the neuron as a bidirectional hub, the source and the product of neural functions. Accounting for a neuron's bidirectionality introduces a certain degree of equivocality, indeterminacy, and randomness into our understanding of neural function. I argue that while this randomness can be reduced for specific, well-defined functions in a particular individual using a top-down approach, it persists for the larger task of determining the neural structures and activity patterns that contribute to consciousness and specific psychological functions. The methodological implications of this shift in perspective include a radical developmental perspective, the need for interdisciplinary frameworks, a focus on cross-level translations and function-specific analyses, and the importance of individual case studies. Here, Developmental Embodiment Research provides a framework that allows us to study neurons in their capacity as bidirectional hubs within a multilevel pathway model of psychobiological development, and in their double role as both products and producers of neural function. For the disciplines, considering neurons in context ultimately demands reorienting the relationship between neuroscience and psychology in their shared endeavor to unveil the material basis of consciousness. Analyzing the neuron in context represents both new obstacles and fresh perspectives for theoretical psychology.

Vanessa Lux, PhD, is a research fellow at the Department of Genetic Psychology, Faculty of Psychology, Ruhr-Universität Bochum, Germany. Her research interests include the impact of stress on psychobiological development and underlying neuroepigenetic mechanisms from the perspective of Developmental Embodiment Research. Emphasizing an interdisciplinary perspective, she is also concerned with the social and cultural implications of neuroscience and epigenetics as well as the integration of different data types used to describe human behavior and experiences, including high-dimensional bio-data and digital behavioral data.

Bochum, Germany   Vanessa Lux

Acknowledgments

This book joins the journey of millions of other books: it was long in the making. Along the way, friends and colleagues supported and inspired my journey, of which only a few can be named. Most importantly, I need to mention my family, who generously supported me with their love and patience during the writing process. Without them, this would not have been possible. The first rough idea for the book stemmed from intense conversations with Jaan Valsiner, whose works strongly influenced my theoretical focus on development and bidirectionality. I also thank Robert Kumsta for the opportunity to join the Genetic Psychology Group at Ruhr-Universität Bochum as well as his ongoing support and openness for these conceptual discussions. The book builds on more theoretical pillars and works of others than the references reflect. Most importantly, the general thinking was strongly inspired by Sigrid Weigel and the discussions at the Center for Literary and Cultural Research (ZfL) Berlin during her leadership. It also builds on a specific perspective on the biological basis of psychological functions, which follows the works of Leontiev and early debates on the relationship between evolutionary theory and developmental biology adapted by Klaus Holzkamp, Ute Osterkamp, and others as the founding principles of German Critical Psychology. My constant struggle with the relationship between psychology and biology, from evolutionary theory to genetics and neuroscience, I share with Wolfgang Maiers. The methodological perspectives throughout the book on the relationship between different types of data and psychological theory have been inspired by the works and discussions with Morus Markard. Furthermore, I would like to thank Suparna Choudhury for introducing me to the concept of bio-looping and for her support during a research stay at McGill University, during which I dived into the origins of Hebb's notion of plasticity, and Melanie Krüger for her contributions to the Developmental Embodiment Research Framework, which strongly inspired this work. Finally, I would like to thank the editorial and production team at Springer Nature for their support as well as Jaan Valsiner in his role as series editor for helpful comments on an earlier draft of this book.

Contents

Chapter 1
From the Perspective of a Neuron

Neuron—A late nineteenth century Greek term, refers to highly specialized "nerve cells." A neuron exhibits a highly complex repertoire of specialized membranous structures, embedded ion channels, second messengers, genetic and epigenetic elements and unique complements of various proteins such as the receptors.

Neurons are excitable cells (i.e., able to conduct electrical impulses of action potentials), which form elaborate networks through axons and dendrites. This ensemble is responsible for integrating, processing and transmitting information, and forms the basis for e.g., coordinated muscle movements and brain functions, including learning and memory formation. (Binder et al., 2009, p. 2751)

Neurons are exciting. First and foremost, because they are excitable cells. As the basic elements of the nervous system, they transmit information across the body through electrical and chemical signals. Second, neurons connect with other neurons to form neural networks, and these networks are important parts of the biological structure that enables us to move, sense, feel, and think. Damage to these neural networks results in functional impairment. A neuron's characteristics as an anatomic and functional unit, its abilities, and its limits constitute its contribution to these networks. Third, psychobiological development depends on the development of single neurons. We do not know how many neurons need to function before the first movements and sensations are possible or how complete their development needs to be. Clearly, their functionality precedes any psychological function. Thus, we need to understand the neuron and its characteristics to understand the biological foundations of the mind.

Since its first description more than a century ago, physiology and neuroscience have provided tremendous insights into the workings of this excitable cell. However, its functionality in vivo remains astonishing to this day. Specifically, how neurons interconnect with each other and other brain cells, how they work together within these networks, and how our mental capacities rely on the functionality of these networks are still not entirely understood. At the same time, framing the single neuron as the ultimate source of mental functions has continuously spurred criticism from scholars in the fields of psychology, pedagogy, and philosophy of mind. They reject the underlying notion of neuroreductionism, which often parallels subjective experiences with simple muscle contractions.

V. Lux, *The Neuron in Context*, SpringerBriefs in Psychology, https://doi.org/10.1007/978-3-031-55229-8_1

At the core of this tension is our theoretical understanding of the neuron, which, in its classical version of the neuron doctrine, singles out the neuron as an individual cell. This not only stands in stark contrast to the growing consensus in neuroscience that neural networks are the actual appropriate unit to match neural structure to function. It also conceptually stands in the way of newer attempts to systematically account for the bodily conditions of our mental functions (e.g., in embodied cognition approaches). To avoid the neuroreductionism trap, these approaches postulate an unresolvable incongruency between our subjective experiences and the underlying (neuro)physiological structure. Most importantly, our current neuron theory does not match the growing knowledge about this exciting cell, which shows that the neuron depends on its context to function properly and that its contributions to specific neural networks, let alone psychological functions, are not as fixed as we like to picture. Therefore, this is a call for a new neuron theory. One that pictures the neuron in context. One that helps us understand why neurons are unique and at the same time erratic in their contribution to mental functions.

Constructing the Single Neuron

Our current understanding of the neuron as a functional unit of the nervous system emerged in the late 1880s and the 1890s, following the development of the Golgi staining method. Invented by Camillo Golgi,[1] this silver impregnation marks only a few neurons within a brain section but highlights them in great detail, including their axons and dendritic branches (see Fig. 1.1). Singled out this way, the technique made visible that the nervous system does not represent a connected tissue mesh (*reticular theory*) but consists of distinct cell entities (*neuron theory*). This marked a paradigm shift, to which, among others, Santiago Ramón y Cajal (1893, 1900), Wilhelm His (1889/2017), Albert von Kölliker (1890), Gustaf Retzius (1890), and Arthur van Gebuchten (1891) contributed detailed empirical observations at the time. In 1891, Wilhelm Waldeyer propagated their findings in a lecture at a meeting

---

[1] Camillo Golgi (1843–1926) was an Italian biologist and pathologist who worked intensely on the tissue and cell composition of the central nervous system. He described several cellular and brain structures and discovered, among others, the Golgi apparatus, the Golgi tendon organ, and the Golgi tendon reflex. For his Golgi staining method, he first immersed the nervous tissue in potassium dichromate for several month to harden it and then stained it with diluted silver nitrate for several days. This technique would make a small number of neurons with their cell bodies, axons, and dendrites visible randomly across the tissue (see Fig. 1.1). Golgi's staining method, although work intensive, was far superior to other staining methods at the time, which did not work as well in nervous tissue. With this technique, Golgi was able to depict for the first time the different shapes of neurons in different brain regions, such as the olfactory bulb, the neocortex, the hippocampus, and the cerebellum (Bentivoglio et al., 2019). In 1906, he was awarded the Nobel Prize in Physiology or Medicine together with Santiago Ramón y Cajal for his work on the nervous system. Despite his own staining method showing the cellular structure of the nervous system, Golgi was a strong protagonist of the reticular theory throughout his lifetime (Cimino, 1999).

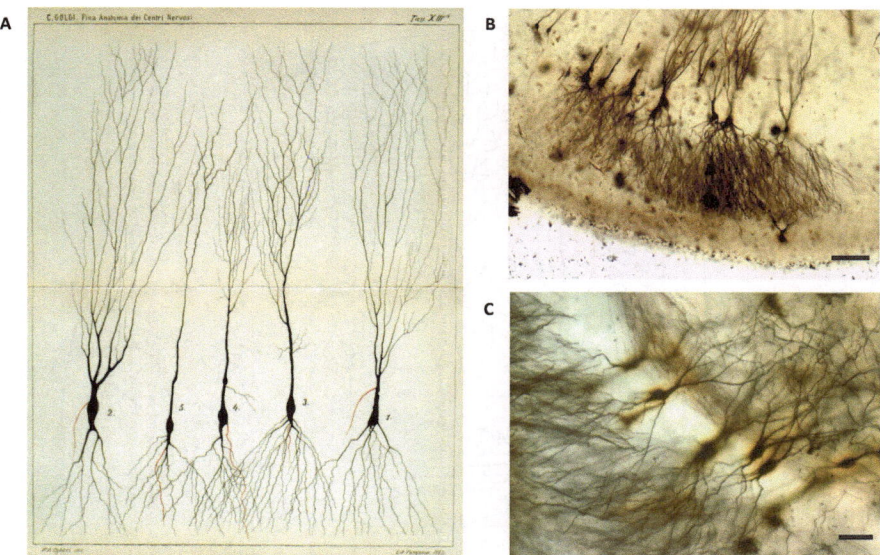

Fig. 1.1 Drawing and images of Golgi-impregnated hippocampal neurons of the rabbit brain. (**a**) Some ganglion cell types in the convoluted gray layer of the pes Hippocampi major. Plate XIII from Golgi (1885, p. Appendix: Tavola XIII). In the plate description, Golgi emphasized the different shapes of the neurons 1–5 depicted in the drawing. (**b, c**) Picture of Golgi's original slides with impregnated neurons. Scale bars: (**b**) 200 μm, (**c**) 50 μm. (Adapted from Bentivoglio et al. 2019, CC BY 4.0 Deed)

of the *Berliner medicinische Gesellschaft*. In his report, he coined the term "neuron" and summarized the empirical evidence in a "basic law" (*allgemeines Grundgesetz*). According to this basic law, neurons represent independent anatomic, trophic, functional, and developmental units that transmit electrophysiological impulses from the central nervous system to the muscles and vice versa. Waldeyer states:

> The nervous system consists of numerous nerve units (neurons) that are neither anatomically nor genetically connected with one another. Each nerve unit is composed of three parts: the nerve cell, the nerve fiber, and the fiber arborization (terminal arborization). The path of physiological conduction can go in the direction from the cell to the fiber arborization or in the reverse direction. The motor conduction occurs only in the direction from the cell to the fiber arborization, the sensory now in the one, now in the other direction. (Waldeyer, 1891, p. 52, see Fig. 1.2)[2]

---

[2] "Das Nervensystem besteht aus zahlreichen untereinander anatomisch wie genetisch nicht zusammenhängenden Nerveneinheiten (Neuronen). Jede Nerveneinheit setzt sich zusammen aus drei Stücken: der Nervenzelle, der Nervenfaser und dem Faserbäumchen (Endbäumchen). Der physiologische Leitungsvorgang kann sowohl in der Richtung von der Zelle zum Faserbäumchen als auch umgekehrt verlaufen. Die motorischen Leitungen verlaufen nur in der Richtung von der Zelle zum Faserbäumchen, die sensiblen bald in der einen, bald in der anderen Richtung" (Waldeyer, 1891, p. 52, English translation: Jones, 1994, p. 4). The term "genetic" is used here in the sense of

Fig. 1.2 Drawing of different types of single neurons in the mammalian cerebral cortex by Wilhelm Waldeyer. In his description of the drawing, Waldeyer also mentions that Golgi assumed a direct connection of the neurons with blood vessels and the connective tissue via their complex arborizations, while he emphasizes that Ramón y Cajal and Retzius did not find any evidence for this assumption. (Adapted from Waldeyer, 1891, p. 35/Fig. 7)

Later, Ramón y Cajal (1906) and Charles Scott Sherrington (1906) added the notion of neurons as polarized units, with the axon as transmission line and the dendrites as receptor lines, as well as the anatomical description and interpretation of synapses as asymmetrical, one-directional transmission hubs to this list (Guillery, 2005). This collection of neuron characteristics became known as the "neuron doctrine" (e.g., Fodstad, 2001; Guillery, 2007). It built a powerful template for the analysis of neural circuits and led to the successful characterization of neurons and their functions all the way down to the molecular level (Kandel, 2000; Cowan & Kandel, 2001).

The paradigmatic strength of this focus on the single neuron and its characteristics spurred by the neuron doctrine is well documented in two passages of Eric R. Kandel's Nobel Lecture.[3] In his recollection, he explained that for the purpose of studying learning and memory, they focused on single neuron properties in the hippocampus:

"developmental," while in the following, if not otherwise stated, I use the term to refer to the DNA sequence.

[3] Kandel received the Nobel Prize in Physiology or Medicine in 2000 together with Arvid Carlsson and Paul Greengard for their findings revealing important parts of the molecular mechanisms involved in signal transduction between neurons specifically at the synapses.

> We were initially interested in a simple question: Are the electrophysiological properties of the hippocampus, which were thought to be the key hippocampal cells involved in memory storage, fundamentally different from other neurons in the brain, such as the well-studied motor neurons in the spinal cord involved in simple movement? In the course of studying the pyramidal cells of the hippocampus, it became clear to us that all nerve cells have similar signaling properties. (Kandel, 2000, p. 393)

As the signaling properties of the individual neurons did not explain the memory storage function of the hippocampus, Kandel and his colleagues then went on to study how single neurons act within a stable neural network and how these interactions are influenced by learning. For their studies, they used the California seasnail Aplysia, which is known for its very few, but large neurons. Choosing this model organism allowed them to trace molecular changes at the single neuron level with the methods available at the time:

> In addition to being few in number, these cells are the largest nerve cells in the animal kingdom, reaching up to 1,000 μm in diameter, large enough to be seen with the naked eye [...]. Because of their extraordinary size and their distinct pigmentation, it is possible to recognize many of the cells as unique individuals. One can record from these large cells for many hours without any difficulty, and the same cell can be returned to and recorded from over a period of days. The cells can easily be dissected out for biochemical studies, so that from a single cell one can obtain sufficient mRNA to make a cDNA library. Finally, these identified cells can readily be injected with labeled compounds, antibodies, or genetic construct procedures which opened up the molecular study of signal transduction within individual nerve cells. (Kandel, 2000, pp. 395–396)

This second passage shows that the focus on single neurons facilitated a pragmatic experimental reductionism, which ultimately led to the discovery of the molecular mechanisms involved in synaptic signal transmission for which Carlsson, Greenblatt, and Kandel received the Nobel Prize. In this sense, the South African Neuroscientist and founding editor of the *Annual Review of Neuroscience*, Willam Maxwell Cowan and Kandel reflect together that "the roots of modern neuroscience reach back to the latter part of the nineteenth and early years of the twentieth centuries when the concept of the neuron as the cellular basis of all nervous systems was established and the essential features of nerve cell signaling and synaptic transmission were laid down, together with the emerging insight of the fundamental role played by neuron-neuron interactions in the integrative action of the nervous system" (Cowan & Kandel, 2001, p. 595). Thus, singling out the neuron within the nervous tissue became an important starting point for modern neuroscience and the way we search for the neural basis of mental functions.

Between Networks and Neuroreductionism

Although our knowledge about neuronal function greatly advanced ever since, this focus on the neuron as the functional unit of the nervous system, and especially as the functional unit of the brain, still shapes our view of the mind-brain relationship. "As such there is the basic assumption that findings in single cells and small cell assemblies translate not only into local but also into far-ranging, large-scale net-works, which finally help to explain higher cognitive functions" (Schmidt-Wilcke et al., 2018, p. 8). This assumption also underlies computer simulation programs, such as GENESIS or NEURON, which model neuronal networks in silico (Bower & Beeman, 1998; Hines & Carnevale, 2001). However, it does not stand undis-puted–on the contrary. Network theories such as the concept of small-world net-works (Bassett & Bullmore, 2016) and phenomena such as neuronal oscillation and synchronicity (Singer, 1999), the study of resting state activity, and the rising con-nectivity paradigm (Raichle, 2009; Sporns, 2013; Sporns et al., 2005)–all of these booming fields of neuroscience challenge the atomic perspective of the neuron doc-trine. Instead, attention is growing for integrative modeling of neural processes (Miller & Cohen, 2001) and dynamic system approaches to brain function (Friston et al., 2014; McKenna et al., 1994; Singer, 2013; Sussillo, 2014). Even at the micro level, phenomena such as collective electric signaling of groups of neurons in the brain (Xu et al., 2018), the two-way signaling of electrical synapses, and the double role of dendrites in receiving and transmitting information (Levitan & Kaczmarek, 2015) question fundamental assumptions of the traditional version of the doctrine. At the single-cell level, multiple layers of neural activity with potentially indepen-dent functions associated with different cellular sites, such as dendrites, axons, syn-apses, and cell bodies, have been reported (for a concise overview, see Sidiropoulou et al., 2006). These findings indicate that nearly every core assumption of the neu-ron doctrine needs revision, and the question arises as to whether the doctrine still holds considering the growing evidence for the embeddedness, equipotentiality, and context-dependence of neuronal function. Consequently, a group of leading neuro-scientists stated in 2005 in an article published in *Science* that

> the complexity of the human brain and likely other regions of the nervous system derive from some organizational features that make use of the permutations of scores of integrative variables and thousands or millions of connectivity variables [...] and perhaps integrative emergents yet to be discovered. The answers extend well beyond explanation by the neuron acting as a single functional unit. (Bullock et al., 2005, pp. 792–793)

Others followed, and the field turned increasingly to "systems neuroscience" (van Hemmen & Sejnowski, 2006) and "integrative neuroscience" (Schmidt-Wilcke et al., 2018, p. 8). These approaches point out the "current reductionist approaches," which are not sufficient to understand how brains and organisms operate in "natural environments" (Wagner & Gaese, 2006, p. 23): "Because organisms are composed of many molecules and cells, we would not understand how these interact to create something like consciousness even if we had worked out the functioning of all chan-nels, cellular propagation mechanisms, and synapses" (Wagner & Gaese, 2006, p. 22). These approaches criticize the limits of "microscopic scale neuroscience," focusing on single cells or parts of cells in the context of specific brain functions, and emphasize the need to "bring together models of the brain within and across scale" (Gordon, 2003, p. S2). They argue for the recording and use of "cortical states," small neural population activity, or other types of network measures, instead of single neurons, when linking neural activity to function (Kenet et al., 2006;

Schöner, 2020). It appears that the neuron vanishes within this systemic context. Did we already enter the "post-neuronist era" (Guillery, 2005, p. 1281) of neuroscience?

The shift away from the cellular entity of the neuron to the network level clearly reintroduced a blurriness in our image of the brain's function, which is difficult to endure. For decades, the neuron doctrine shaped our view of the nervous system in a way that allowed the rigid application of experimental methods. It seemed only a matter of time and computational capacity to solve the puzzle of consciousness. This came at the prize of a reductionist perspective that put the single neuron as the starting point of our experiences and behavior and ignored its bodily and developmental preconditions. Not long ago, this perspective dominated popular perceptions of neuroscience and its achievements. One prominent example of this simple version of neuroreductionism is the interpretation of the Libet experiments as a test of "free will" (Haggard & Libet, 2001; Libet, 1999). In a series of experiments, Benjamin Libet demonstrated that the readiness potential of motor neurons detected by electroencephalography (EEG) precedes the time point at which a person consciously recollects the decision to move their finger by about 500–800 ms (Libet et al., 1983). These findings were later replicated using a functional imaging approach (Soon et al., 2008). Libet (1999) and other neuroscientists (Prinz, 2008; Roth, 2003; Singer, 2004) interpreted the findings as proof that our conscious decisions are predetermined by their underlying neural activity, and thus, our subjective experience of "free will" just an "illusion." This sparked several controversies regarding the role of consciousness in behavior (Pockett et al., 2006), the accountability of actions and decision-making (Geyer, 2004), and the implications for ethics and the law (Kolber, 2016; Pauen, 2007). The notion that our behavior and mental acts are predetermined by their underlying neural activity not only contradicts our modern subjective experience as voluntarily acting individuals. It also closes the epistemological gap between these different levels of description, which, for the last 100 years, was traditionally occupied by psychological theory.

As a consequence, psychologists felt under pressure from neuroscience to defend the validity of their discipline and the way they study the workings of the mind (Fiedler et al., 2005). The debate gave rise to alternative explanations for Libet's experiments. For example, Wolfgang Maiers (2009) points out that the neuroreductionism in the debate results from a conceptual fuzziness at play: the basic mechanisms at the neuronal level, which enable simple motor actions, are mistaken for those underlying the complex spectrum of human consciousness. Maiers argues that what is defined as a voluntary act in the Libet experiments–the movement of the finger–is only the last isolated step in a series of preparation practices that include agreeing to participate in the experiment, coming to the lab, sitting down on a chair, waiting for the start signal by the experimenter, and concentrating on the task. This preparation phase, related to different action potentials at all these different steps and their experience as one entity, is core to human experiences in everyday life. Excluding these features of our experience results in a theoretical shortcoming, according to Maiers: "The first person experience to act voluntarily, the fiat of 'I will' would, then, become only the secondary personal awareness of a successfully realized process of neuronal action planning [...]. Libet's instantaneous choice is at best the last step in a process set in long before" (Maiers, 2009, p. 110). Taking this last step as the actual voluntary act means "[r]eplacing the agents by their natural (bio-social) organs of agency" (Maiers, 2009, p. 110)—the person by its brain. Philosophical accounts of the "free will" terminology point to additional conceptual ambivalences, and alternative models of consciousness, integrating the findings within a multi-level concept of personhood, action responsibility, and decision making, were proposed (for example, Murphy et al., 2009; Pauen, 2007; Pockett et al., 2006).

As the example of this debate shows, the notion of the single neuron as direct material basis and source of consciousness favors reductionist explanations of human behavior and mental functions if the broader context of the neuron's activity is not considered. This contextualization implies different levels of explanation, from biophysical mechanisms to subjective experiences, which are often framed as irreducible perspectives on the same phenomenon, producing an epistemological gap at the core of the body-mind problem. As mentioned above, this gap has traditionally been the place of psychological theory since the days of Freud, Wundt, and Pavlov, and their attempts to connect behavior, sensations, and introspective accounts with what was known about the biology of our central nervous system at their time. More recently, phenomenological accounts of consciousness and embodiment theories have taken up the notion of contextuality. These emphasize in an even more radical way that the foundation of consciousness lies not only in our neurons or the brain, but that the mind is an activity of the living body, integrating our bodily experiences in interaction with the environment (T. Fuchs, 2018, 2020). In its consequences, this perspective views the mind as depending on and being situated also in our skeletal structures, our bones, muscles, and organs (Pfeifer et al., 2007), our intersubjective encounters, and even our clothes, houses, computers, thus, our entire cultural surroundings, our ecological niche, representing the mind's collective materiality (Varela et al., 2016). From an epistemological standpoint, the single neuron loses here its importance and centrality, vanishing within complex neural networks and their multilayer ecological niche. One could even argue that the neuron gets lost within this material context.

The Task: Conceptualizing the Embedded Neuron

On a theoretical level, the contextuality and embeddedness of the neuron and neuronal function are widely acknowledged, although not in the far-reaching sense proposed by phenomenological and embodiment theories. That "cellular and molecular neurobiology do not exist in a vacuum" (Levitan & Kaczmarek, 2015, p. 3) is undisputed. However, how this context is related to neural function is, for most scenarios, unknown. In addition, the notion of embeddedness or contextuality of neural functions does not translate into research practice. Most theoretical, computational, and experimental attempts that go beyond mere theoretical reflections have dealt with this embeddedness of the neuron in the same way, investigating how the abilities and constraints of a single neuron translate into the systemic processes underlying mental functions. We also observe this in recent attempts to characterize different types of neurons and other brain cells based on their transcriptomes[4] (Darmanis et al., 2015; Hawrylycz et al., 2012; Peng et al., 2021). The hope is that the heterogeneity, for example, at the transcriptome level, translates into and thus explains the functional heterogeneity at other cellular and functional levels. Contrary to expectations, these attempts revealed, for example, that neurons with similar transcriptomes show tremendous morphological variety (Peng et al., 2021). Despite the growing knowledge of this complex cellular composition of neural networks and brain tissue, which form the direct environment of the single neuron, the vision is still to complete the puzzle by putting together all the single pieces (Koch, 2019; Peng et al., 2021). In this scenario, the neuron is still considered the origin and ultimate source of our ability to move, sense, feel, and think. Or, as Koch once put it as a question for the "good fairy godmother of science (GOOFGOOS)": "are there specific neurons whose activity mediates consciousness?" (1996, p. 492). However, even if we hold on to this notion, it is only one side of the story. What is missing is the question of how neurons are affected by this dynamic connectedness, embeddedness, and embodiment. What does it mean for a single neuron, for its development and functionality, to be one among many in an intricate network of other neurons, body parts, and the environment that brings about our mental functions? In addition, what does the embeddedness of the single neuron mean for our understanding of the relationship between mental functions and their underlying neural structures? How can this knowledge be integrated into our understanding of single neurons? Clearly, it is time for a new neuron theory, one that also tells this other side of the story.

What Makes a Neuron Unique

For this, we first need to understand what constitutes the single Neuron, and what makes it unique compared to other neurons. From the very beginning, neurons in the spinal cord were categorized by their function as sensory neurons, motor neurons, and interneurons. According to this schema, sensory neurons are specialized to signal from sensory cells to other neurons in the central nervous system (CNS). They are activated by physical or chemical input based on our senses (e.g., smell, sound, light, touch, and temperature). Motor neurons signal from the CNS to muscles, organs, or glands, and are involved in controlling all muscle-based body

---

[4] The transcriptome is the full range of messenger RNA (mRNA) transcribed by an organism, cell type, or tissue depending on the unit of analysis. For neuron classification, for example, the transcriptome of single neurons or glia cells is analyzed separately, and cells with similar transcriptomes are grouped together to subtypes. This grouping step is usually done using unsupervised learning methods such as principal component analysis or cluster analysis.

functions including digestion and the heartbeat. Interneurons connect sensory and motor neurons but can also connect with each other to form complex neural networks. This neat classification does not hold true for neurons in the brain. The question "How many different neurons are there in the brain?" is still unanswered (Sharpee, 2014). Although some clearly exhibit functions that would characterize them as sensory or motor neurons, most neurons in the brain develop multiple connections with different types of neurons. They form small networks of neighboring cells or connect neuron groups across long distances or both. In addition, they occur in all sizes and shapes, with morphological variations in axon length, degree of myelination, and different numbers of dendrites and spines. They differ in their synaptic properties and connect with their synapses to the cell bodies or to the dendrites of other neurons.

One system to further characterize neurons in the brain is their classification by neurotransmitters, such as acetylcholine, glutamate, dopamine, serotonin, noradrenalin, and histamine. This classification assumes that neurons that use the same neurotransmitter build functional units or networks because they can communicate with each other on a chemical basis. However, within each group of neurotransmitters, we find tremendous heterogeneity regarding all other neuron properties, including electrical properties, cell shapes, gene expression profiles, projection patterns, synaptic properties, and the type of input received. Subclassifications along these properties failed (Mott & Dingledine, 2003). Also, dual- and multi-transmitter neurons were detected, which are able to release more than one neurotransmitter even at the same time (Vaaga et al., 2014). Other classification attempts focusing on gene expression patterns and phylogenetic approaches provided further evidence for the tremendous diversity of neurons across cellular properties, activation states, brain areas, and developmental stages (McKenzie et al., 2018; Moroz, 2018, 2021; Naumova et al., 2013; Negi & Guda, 2017). Thus, recent efforts, again, attempt to classify neurons primarily according to their function, that is, under which circumstances they are activated or not activated. These efforts make use of the available large-scale connectivity datasets (Bota & Swanson, 2007; Sharpee, 2014). The vision is to identify sets of neurons that fulfill the same function, for example, fire together based on a specific sensory input, are involved in the same motor function, or contribute to the same memory mechanism. Some of these efforts have been successful, especially for sensory neurons specialized for a particular sensory stimulus or parts of a stimulus (see, e.g., Emery et al., 2016; Enander & Jörntell, 2019; Millman et al., 2020; Mountcastle, 1957). Other findings further complicate the functional classification of neurons. For example, the discovery of "mirror neurons" showed that some (but not all) motor neurons underlying a certain action also participate in the perception and inner representation of the same action (Gallese, 2013; Kohler et al., 2002; Rizzolatti & Craighero, 2004; Rizzolatti & Fogassi, 2014; Rizzolatti & Sinigaglia, 2016). Also, some of these "mirror neurons" are highly specialized and can be further differentiated according to the sensorial properties of the perceived action, while other "mirror neurons" are activated by all signals related to the action (Kohler et al., 2002). Overall, this points towards a fascinating parallelism of highly functional specialization and, at the same time, a broad multifunctional capacity of neurons.

Even the last bastion of a neuron-based functional classification, the somatotopic map, known as Penfield's homunculus, has been shattered recently. Wilder Penfield, Edwin Boldrey, and Theodor Rasmussen established the idea that neural projections from the sensory neurons of different body parts are fixed during development, resulting in stable somatotopic maps in the somatosensory cortex (Penfield & Boldrey, 1937; Penfield & Rasmussen, 1950). Nowadays, functional magnetic resonance imaging (fMRI) studies demonstrated the existence of multiple homunculi in the brain, located not only in the cortex but also in subcortical structures (Zeharia et al., 2012, 2015). In addition, Tal et al. showed that the homunculus in the somatosensory cortex is not produced by neural activation alone, but also by parallel functional inhibition, which sharpens the body representations of these maps (Tal et al., 2017). This means that the neurons are not only activated by sensory input from the body part they specialize for but also inhibited by sensory input from other body parts, already representing at least a bidirectional responsiveness. To put it simply, tongue-related neural networks fire when the tongue moves or is touched and, at the same time, block neuronal networks related to other body parts–two functions that differ strongly in their functional polarity and target area. This finding is likely to translate to other regions and homunculi. It has long been acknowledged that the constitution of a homunculus depends on experience. However, the dynamics of its organization and dependence on functional connectivity, as indicated by these studies, afford a much higher degree of plasticity and flexibility of the underlying neural networks, specifically at the single neuron level, than previously assumed. The findings point toward an operation-based concept of somatotopic organization in the brain based on semi-flexible functional units instead of anatomically fixed maps. Thus, although we find these maps in current activation states, from the perspective of the single neuron, participating in such a map is not as constitutive for the neuron's identity as we thought. These findings indicate that somatotopic representations are a general organization and processing principle across the brain, recruiting neurons as needed instead of gluing them to the sensory or motor action pathway defined by these maps. Consequently, somatotopic maps turned out to be not as useful as hoped to classify single neurons and small groups of neurons according to their function, even in somatosensory areas.

Several additional observations of neuron properties reflect this multiplicity of functions at the single neuron level. Parallel local and distant connections of single neurons (for example Dahmen et al., 2022; E. C. Fuchs et al., 2016; Sepulcre et al., 2010), parallel electrical and chemical transmission systems (e.g., Faber & Pereda, 2018; Hormuzdi et al., 2004; Levitan & Kaczmarek, 2015, pp. 41–62), and different transmission cycles and rhythms (e.g., Auksztulewicz et al., 2019; Doi & Kumagai, 2005; Fries, 2015; Ibarra-Lecue et al., 2022; ten Oever & Sack, 2019) contribute to the complex functional connectivity of individual neurons.

Glia Cells–The Rising Stars

The real-life picture is even more complicated than this methodological focus on the network level and neural connectivity suggests. The neuron does not perform its functions alone as a single cell but is embedded within a complex network of other cells. This includes fatty cells and other cell types within the cellular connective tissue, and most importantly, glia cells. Neurons are always connected to glia cells. These provide a supporting cell structure, which holds the neurons in place and insulates them from each other. This supporting structure also participates in larger morphological changes by facilitating and coordinating neuron migration, especially during cell development (Rakic, 1981). In the brain, glia cells coordinate neuron migration across different brain layers. In addition, glia cells contribute to the cell metabolism of neurons by providing nutrition and oxygen and help maintain the synaptic properties of neurons (Allen & Lyons, 2018; De Backer & Grunwald Kadow, 2022; Stevens, 2003). Furthermore, they protect neurons by destroying pathogens and removing dead neurons (Norris et al., 2018; Rock et al., 2004). Most importantly, certain glia cells provide myelination. The connection between a neuron and its myelinating glia cells is functionally highly relevant and has a symbiotic character. Myelination enables the formation and transmission of action potentials along the axon. In the brain, myelination is provided by oligodendrocytes. One oligodendrocyte myelinates only parts of the axon of a neuron; therefore, several oligodendrocytes are needed for one axon, but each oligodendrocyte may participate in the myelination of several axons of different neurons. The relationship between oligodendrocytes and neurons has long been considered external. However, it is now known that they exchange cell vesicles and communicate via chemical signaling with each other (Ahmad et al., 2022; Campbell, 2003; Frühbeis et al., 2013). The same seems to be the case for astrocytes and microglia, and first studies indicate a functional role for glia cells in connectivity and signal transmission (Fields et al., 2015; Fields & Stevens-Graham, 2002; Kahali et al., 2021).

Thus, the function of a neuron is the product of the function of a whole group of cells, including other neurons, as well as different types of glia cells. The unique contribution of each cell is still not fully understood. If we consider this seriously, it contradicts the function-based classification of *single* neurons. Moreover, we would need to model neuronal function as a summary of these different cell functions, or the field gradient of a specific locally defined part of tissue in a certain state of activation. In addition, a growing body of research investigating the role of glia cells in mental functions (Elsayed & Magistretti, 2015; Nagelhus et al., 2013; Yamamuro et al., 2015)–already coined "[t]he glia doctrine" (Nagelhus et al., 2013, p. 449)–questions the focus on neuronal connections still dominating the connectivity paradigm. Reconciling both perspectives is one of the most important tasks in this field. The concept of the neuron and how its functions are produced are at the core of this task. New detection methods that record different layers of neural activity at the single-cell and network level as well as at the molecular level, including multiple cell recordings, optogenetics, gene expression analysis, and proteomics studies, and their integration through computational modeling, provide the means for this endeavor (Altimus et al., 2020).

Neuroepigenetics is one of the fields that emerged from these technical advances. As I will argue in the following chapters, neuroepigenetic information has the potential to specifically characterize the neuron, as it develops in response to its spatial and temporal context, even beyond current activation states. By enabling us to register parts of a neuron's developmental and activation history at the molecular level, neuroepigenetics provides insights into another layer of neuron diversity, further emphasizing the context dependence of the single neuron and its function.

Molecular Epigenetics Paving the Path

The power of the Golgi staining method was its reductionist output, making the single neuron visible within the cell mess of whole brain tissue. Every approach that followed to visualize or otherwise detect the properties of single neurons and other brain cells also enacted some methodological reductionism. Often, these techniques exclude one another. The advancement of knowledge regarding the diversity of neurons and glia cells in the brain is largely due to the advancement of alternative detection methods. Most of the previously discussed properties of the neuron are situated at morphological and neurophysiological levels. They are measured and recorded using physiological and biochemical methods such as electrophysiology, imaging approaches, and tissue staining (Backer, 2022; Martin, 2019; Xiong & Gendelman, 2014). Gene expression analysis of the brain transcriptome at the level of brain areas (Negi & Guda, 2017; Sjöstedt et al., 2020) and at the single-cell level (Darmanis et al., 2015), and its combination with microarray-based immunohistochemistry to measure the proteome of the human brain (Uhlén et al., 2015), provided a further detailed and partially divergent typology of human brain cells. Efforts to combine these different identification and characterization methods are at the heart of the current debate regarding the nomenclature of brain cells (Yuste et al., 2020). In this context, the study of molecular epigenetic mechanisms in brain cells and their potential use as biomarkers for single neuron activity and overall neural function reveals additional layers of neuron diversity. The study of molecular epigenetic mechanisms in brain cells picked up speed after several key findings indicated the previously overlooked changeability and environmental sensitivity of these markers in differentiated body cells (Box 1.1).

Molecular epigenetics is defined as "the study of mitotically and/or meiotically heritable changes in gene function that cannot be explained by changes in DNA sequence" (Riggs et al., 1996, p. 1). The three most commonly studied epigenetic mechanisms are variations in DNA methylation, different types of histone modifications, and RNA interference.

Box 1.1 Molecular Epigenetic Mechanisms
The current understanding of molecular epigenetic mechanisms includes DNA methylation, histone modifications, and RNA interference.

DNA methylation is the covalent transfer of a methyl group to the C-5 position of a cytosine ring in the DNA via DNA methyltransferases (DNMTs). In the mammalian genome, most DNA methylation in somatic cells occurs at the cytosine-phosphate-guanine (CpG) sites. CpG sites are often clustered in CpG islands in promotor regions of expressed genetic loci, giving rise to the understanding that DNA methylation regulates gene expression by enhancing or inhibiting transcription factor binding. There are two types of DNMTs: One type (DNMT1) preferentially methylates hemi-methylated DNA, ensuring the replication of DNA methylation patterns during mitosis, and the other type (primarily DNMT3-A,B) functions as a de novo DNMT, preferentially methylating prior unmethylated sites within the genome. Nearly all DNA methylation is removed during zygote formation and is reestablished by de novo DNMTs during embryogenesis. Observations of dynamic changes in DNA methylation during regular cell cycles in differentiated cells and their sensitivity to environmental signals have initiated the rise of molecular epigenetics as a specific research field.

Histone modifications influence gene expression and DNA maintenance by altering the electrostatic and biochemical properties of histones in eukaryotic cell nuclei. These histones package the DNA into nucleosomes and chromatin fibers and provide an important support structure for the DNA. Several histone modifications have been detected, including methylation, acetylation, phosphorylation, ADP-ribosylation, and ubiquitination, but the functional understanding of most of them is still sparse. In general, it is observed that histone modifications affect the binding of other proteins to the DNA. Histone methylation at arginine or lysine residues of the histones, for example, affects the binding of transcription factors, which enhances or depresses gene transcription at the specific site. Lysine acetylation weakens the electrostatic attraction between neighboring histones, resulting in partial unwinding of the DNA, making it more accessible to transcription factors and other proteins, including DNMTs.

RNA interference is a process in which small pieces of RNA guide the pre- or posttranscriptional silencing of protein translation. Mainly, two types of RNA molecules participate in the posttranscriptional silencing of protein synthesis in the cytoplasm: microRNAs (miRNAs) and small interfering RNAs (siRNAs). miRNAs are derived from noncoding RNA transcripts that fold back on themselves and form short hairpins. siRNAs result from the activity of the enzyme Dicer, which cleaves long double-stranded RNA into shorter double-stranded fragments. The double-stranded miRNA or siRNA fragments unwound; one strand is cleaved by Argonaute proteins and degraded, whereas the other strand is incorporated into the RNA-induced silencing complex (RISC). RISC binds to the target mRNA with a matching mRNA sequence

(continued)

Box 1.1 (continued)

and degrades it, thereby preventing or modifying the translation of the encoded protein in the cytoplasm. The pretranscriptional silencing mechanisms occur in the nucleus. Here, the RNA-induced transcriptional silencing complex catalyzes DNA methylation at genomic positions that complement the siRNA or miRNA sequences. Through these processes, RNAi pathways participate in genome maintenance and the regulation of cell-specific gene expression, especially during development. In addition, they exhibit immune functions, for example, by silencing potentially harmful nucleotide sequences, such as viruses.

These molecular epigenetic mechanisms play an important role in cell development and differentiation. For example, relatively stable DNA methylation patterns across the genome participate in imprinting mechanisms and maintenance of cell type-specific gene expression profiles throughout a cell's life cycle. These methylation patterns contribute to anatomical and functional differences between, for example, epidermal cells and liver cells. In short, they influence a good part of which genes are expressed, and in what quantity. Such mechanisms of gene expression regulation, and especially quantitative fine-tuning mechanisms, play an important role in the establishment and maintenance of cell functions.

For a long time, it was thought that epigenetic regulation of gene expression, most importantly DNA methylation, is stable after cell differentiation and is fully replicated during cell division. This changed when active DNA methyltransferases (DNMT) were discovered that implemented that de novo DNA methylation occurs in differentiated cells (Chédin, 2011; Kato et al., 2007; Okano et al., 1998). In neuronal cells, these de novo DNMTs (DNMT3 a,b) are more strongly expressed than in other somatic cell lines (Feng et al., 2005; Goto et al., 1994). This raised the question of the biological function of post-mitotic changes in DNA methylation. One paradigmatic experiment showed that de novo DNA methylation can be triggered by diet. Waterland and Jirtle (2003) fed yellow (A^{vy}) agouti mice a methyl donor-rich diet and reported that an increased number of their offspring showed a pseudoagouti phenotype that resembled the brown-colored wild type. In addition, they showed that this phenotypic shift corresponded with increased DNA methylation and reduced gene expression of the *agouti* gene. Around that time, first studies also reported that DNA methylation patterns change over the lifespan by showing that older monozygotic twin pairs differ more strongly in their methylation patterns than younger twin pairs (Fraga et al., 2005). Suddenly, DNA methylation patterns seemed not so stable anymore.

In this context, epigenetic mechanisms evolved as possible molecular pathways for the early environmental adaptation of an organism. Physiological programming effects and their potential adaptive and maladaptive developmental impact have been discussed for quite some time in the context of food consumption, lifestyle, and metabolic syndrome (Barker, 2007; Hoffman et al., 2017). With epigenetic mechanisms sensitive to environmental signals, a possible molecular mechanism underlying these physiological programming effects was on the table (Bianco-Miotto et al., 2017; Goyal et al., 2019; Wadhwa et al., 2009). Still, evidence for functionally specific adaptations was lacking. This was until pioneer studies by Meaney et al. in Long-Evans rats showed that maternal behavior-induced changes in the stress response were related to epigenetic modifications at the glucocorticoid receptor gene (*Gr*) in the hippocampus (Labonté et al., 2013; McGowan et al., 2009; Weaver et al., 2004). The *Gr* participates in the regulation of the stress responses at the micro (tissue) and macro (system) levels. At the micro level, it represents an unspecific mechanism that can initiate tissue-specific stress responses. It is expressed in almost every cell in the body and is involved in the activation or repression of various genes. At the macro level, the glucocorticoid receptor is involved in the negative feedback loop of the hypothalamus-pituitary-adrenal (HPA) axis in the pituitary, which regulates the cortisol/corticosterone-based cycle of the stress response. Meaney et al. found that in their rat model, low levels of maternal licking and grooming during early life increased HPA axis responsiveness to stress. This was associated with persistent DNA hypermethylation at specific CpG dinucleo-tides within the hippocampal *Gr* exon 1_7 promoter and increased histone acetyla-tion, facilitating the binding of the transcription factor nerve growth factor-inducible protein A (Ngfia), which increased *Gr* expression (Weaver et al., 2004). Other research groups tried to replicate these findings with mixed results (see Bockmühl et al., 2015; Murgatroyd et al., 2009; for a critical review, see Palma-Gudiel et al., 2015). Some groups found similar mechanisms in other brain regions involved in stress regulation. For example, Bockmühl et al. (2015) investigated epigenetic mod-ifications at the *Gr* promoter region in the paraventricular nucleus (PVN) of the hypothalamus, a region that is directly involved in the regulation of the stress response via the release of corticotrophin-releasing hormone (CRH), which then triggers the release of cortisol in the pituitary. The release of CRH is then reregu-lated by the feedback inhibition effects of cortisol on the pituitary via glucocorticoid receptors. Although Bockmühl et al. found no differences in the proximal *Gr* pro-moter region, they observed hypermethylation at CpG sites in the shore region of a more distal CpG dense island in the *Gr* promoter region. At one of these CpG sites, hypermethylation was robustly maintained for over 3 months. In addition, they reported an increase in overall hypermethylation and age-related increases in *Gr* mRNA transcripts in the PVN of mice that experienced early life stress, indicating a functional role of this hypermethylation of CpG sites in the shore region of the more distal CpG island in *Gr* regulation across the lifespan. In accordance with these findings, epigenetic programming effects affecting the stress response became a number one target for psychiatric genetic studies (Isles, 2015; Isles & Wilkinson, 2008; Ptak & Petronis, 2010). In particular, the effects of early life stress on the epigenome and their potential role in the etiology of psychiatric disorders are under investigation (Bock et al., 2015; Heim & Binder, 2012; Stankiewicz et al., 2013).

Both the liking and grooming behavior-molecular epigenetic fine-tuning com-plex of the stress system as well as the diet-induced DNA hypermethylation in Agouti mice also affected subsequent generations (Waterland & Jirtle, 2003; Weaver et al., 2004). This transgenerational stability supported the notion that these mechanisms were not purely incidental but represented possible environmentally directed adaptive changes. These findings and the term "heritable" fueled the New Lamarckism debate, emphasizing the role of epigenetic modifications as an additional inheritance system (Jablonka & Lamb, 2005; Lux & Richter, 2014; Pigliucci et al., 2010). It also provided a convincing template for potential molecular mechanisms underlying the transgenerational effects of extreme traumata which have been reported in clinical studies of Holocaust survivor families and other trauma survivor groups (Schmidt et al., 2011; Yehuda et al., 2016). However, until now, only a handful of animal studies could show such a transgenerational effect of trauma-like experiences. For example, Isabelle Mansuy and her group at the University of Zurich showed partial transmission of depressive-like behavior and changes in DNA methylation in the brain and sperm induced by early life stress into the third generation via the male germ line (Franklin & Mansuy, 2010). Dias and Ressler (2014) showed the transmission of behavioral and neural correlates of odor fear conditioning into the second generation via the male germline. In human studies, such transgenerational transmission could not be shown despite intense efforts. One study reported changes in DNA methylation at genomic loci encoding FKBP Prolyl Isomerase 5 (Fkbp5), a protein that modulates glucocorticoid receptor activity in response to stress, in the first and second generations of Holocaust survivor families, although in divergent directions and at different CpG sites (Yehuda et al., 2016). These findings were heavily disputed and fueled the already ongoing debate on the transgenerational effects and transmission of trauma, depression, and other mental disorders (Deichmann, 2016; Heard & Martienssen, 2014; Horsthemke, 2018; Yehuda et al., 2018; Yehuda & Lehrner, 2018). Nevertheless, interest in molecular epigenetic correlates for the entire range of psychiatric disorders and mental functions is still growing.

Neuroepigenetics: A New Layer of Neuron Diversity?

In this context, neuroepigenetics emerged as a subfield of molecular epigenetics. Neuroepigenetics is characterized by the application of molecular epigenetic techniques to neurons or brain tissue.[5] Day and Sweatt argued that the heritability criterion of molecular epigenetics, that epigenetic modifications are stable across meiosis or mitosis, is not useful for studies investigating epigenetic modifications in nondividing neurons (Day & Sweatt, 2010). This is despite the fact that among those epigenetic modifications, which have been reported to be transmitted across generations in mammalian models, some have been recorded in brain tissue (Dias & Ressler, 2014; Franklin & Mansuy, 2010). Accordingly, Day and Sweatt define

---

[5] For a conceptual differentiation between the different variants and definitions of epigenetics, see Lux (2013) and Lux and Richter (2014).

"neuroepigenetics as a potential subfield of epigenetics that deals with the unique mechanisms and processes allowing dynamic experience-dependent regulation of the epigenome in nondividing cells of the nervous system, along with the traditionally described developmental epigenetic processes involved in neuronal differentiation and cell-fate determination" (2010; similar Bird, 2007; Sultan & Day, 2011).

This definition includes findings on DNA methylation changes and histone modifications underlying learning and memory, such as those identified by Day and Sweatt in fear conditioning experiments in rats. In a paper discussing the role of epigenetic mechanisms in memory and learning, Day and Sweatt (2011) recapitulated how they set out to examine epigenetic correlates of learning and memory in the first place. They stated that it was the sensibility to environmental signals combined with the stability and self-sustainability of the induced epigenetic changes once they were established, which seemed to make them the perfect target for molecular correlates of memory formation, especially persistent and recurrent memories present in PTSD or anxiety disorders. In particular, DNA methylation seemed to fit this hypothesis. Therefore, when they set out to study epigenetic correlates of fear conditioning in the brain, they searched for stable changes in DNA methylation and corresponding histone modifications. Ultimately, they found DNA methylation changes in their target brain area, the hippocampus, but these changes were not stable (Day & Sweatt, 2010; Sultan & Day, 2011). Instead, at a later time point, more stable DNA methylation changes appeared in cortical areas. Day and Sweatt concluded that these were likely induced by network projections from the hippocampus to cortical areas, possibly mimicking the transformation from short-to long-term memory. For the induction of epigenetic changes via neural projections between brain areas, they coined the term "systems heritability" (Day & Sweatt, 2010, p. 1322). This indicates the dynamic nature of epigenetic mechanisms in brain tissue, which further challenges the traditional view of stable DNA methylation.

In this context, the question arises as to whether there is a functional or mechanistic difference between these neuroepigenetic mechanisms and other molecular cell processes related to the regulation of transcription in different brain cells. Anthony Isles (2015) warns that a broad understanding of epigenetic mechanisms can lead to an unrealistic "hype" of epigenetics in the study of brain development, development of behavior, and psychiatric disorders. A broad definition would not adequately differentiate between transient changes in gene expression related to short-term neural activity and long-term epigenetic modifications of neural network characteristics, which contribute to diverging developmental pathways, for example, in the etiology of psychiatric disorders. However, at the molecular level, it might not be easy to differentiate between short- and long-term epigenetic modifications.

Emerging evidence indicates that epigenetic mechanisms are also involved in the establishment and maintenance of functional sensitivity and molecular feedback mechanisms that regulate synaptic activity (Bayraktar et al., 2020; Gräff & Mansuy, 2008; Sultan & Day, 2011). One potential key function of neuroepigenetic mechanisms could be to coordinate neural activity with the underlying genetic activity across the neuronal cell cycle (Bayraktar et al., 2020, see Lux, 2013, for a detailed discussion). For most biological systems, we must assume that there are backup mechanisms for functionally relevant pathways. Single epigenetic mechanisms are likely involved in more than one gene-regulating cascade, and each functionally relevant mechanism likely builds on several parallel existing cascades. This includes both short-and long-term epigenetic modifications. Thus, in addition to the time course, further characterization of the functional context of epigenetic modifications is required to identify the mechanisms relevant to and involved in divergent developmental pathways and to avoid unrealistic hype.

To account for the functional context of epigenetic mechanisms in neurons, I propose differentiating them into structural-genomic, synaptic, and developmental epigenetic mechanisms (Lux, 2013). *Structural-genomic epigenetic mechanisms* are those that maintain the genomic structure in neuronal cells by being involved in genomic repair mechanisms as well as mechanisms that maintain low expression levels of pro-apoptotic genes and stabilize cell type-specific gene expression patterns. *Synaptic epigenetic mechanisms* regulate gene expression underlying acute synaptic activity, as well as *short-term* adaptations of synaptic plasticity and function. In contrast, *developmental epigenetic mechanisms* contribute to the establishment and consolidation of *long-term* differentiation in structure (neuronal cell type, morphology) and function (synaptic sensitivity, plasticity) of neuronal cells, such as differentiation of neural stem cells, enduring changes in dendrite growth, long-term changes in neuronal plasticity, and long-term changes in transcription levels. In particular, the interplay between environmental signals, such as stress during a sensitive developmental period, and functionally relevant epigenetic modifications in neurons are discussed as potential molecular mechanisms underlying experience-dependent differentiation in psychobiological development and the development of mental disorders (Bock et al., 2015; Heim & Binder, 2012).

If we look at these epigenetic modifications from the perspective of the single neuron, they represent an additional layer of functional specialization. With different timeframes, from short-term transient to long-term stable and even heritable, these epigenetic mechanisms provide a set of molecular tools that allow energy-effective modifications as well as low-energy maintenance of gene expression states for a variety of neuronal functions. However, to understand their role in the single neuron, we need to determine how these epigenetic modifications are implemented, stabilized, and maintained. Important factors influencing these modifications are synaptic activity or neural input, hormonal changes, and the availability and concentration of epigenetic functional enzymes. In the case of RNA interference, disentangling the participating factors is even more complicated, as functionally relevant microRNAs are not only produced by the transcription of noncoding DNA but can also enter the neuron from surrounding cells via exosomes (Chivet et al., 2012; Rajendran et al., 2014). As mentioned above, communication between oligodendrocytes and neurons partly relies on the exchange of such tiny vesicles (Pascual et al., 2019). Thus, a neuron's epigenetic layer results in part from external input produced by or via the cells that surround and connect to a specific neuron under study. Furthermore, neuroepigenetic mechanisms enable neuronal plasticity (Tognini et al., 2015). The dynamic nature of epigenetic mechanisms, including

DNA methylation, may not be unique to neurons. It is, however, more pronounced in neurons, enabling flexible adaptations at the molecular level in these typically nondividing cells. In contrast, neuroepigenetic mechanisms also function as a buffer, protecting neurons from constantly changing environmental signals by providing an energy-efficient way to stabilize neuronal function at the nuclear level. Thus, the question is whether neuroepigenetic mechanisms enable stability or plasticity of function from the perspective of the single neuron. To answer this question, we first need to recapitulate what plasticity means for brain functions as well as for the single neuron.

References

Ahmad, S., Srivastava, R. K., Singh, P., Naik, U. P., & Srivastava, A. K. (2022). Role of extracellular vesicles in glia-neuron intercellular communication. *Frontiers in Molecular Neuroscience, 15*, 844194. https://doi.org/10.3389/fnmol.2022.844194

Allen, N. J., & Lyons, D. A. (2018). Glia as architects of central nervous system formation and function. *Science (New York, N.Y.), 362*(6411), 181. https://doi.org/10.1126/science.aat0473

Altimus, C. M., Marlin, B. J., Charalambakis, N. E., Colón-Rodríguez, A., Glover, E. J., Izbicki, P., Johnson, A., Lourenco, M. V., Makinson, R. A., McQuail, J., Obeso, I., Padilla-Coreano, N., & Wells, M. F. (2020). The next 50 years of neuroscience. *Journal of Neuroscience, 40*(1), 101–106. https://doi.org/10.1523/JNEUROSCI.0744-19.2019

Auksztulewicz, R., Myers, N. E., Schnupp, J. W., & Nobre, A. C. (2019). Rhythmic temporal expectation boosts neural activity by increasing neural gain. *Journal of Neuroscience, 39*(49), 9806–9817. https://doi.org/10.1523/JNEUROSCI.0925-19.2019

Backer, K. C. (2022). Introduction to experimental methods in cognitive neuroscience. In *Mind, cognition, and neuroscience*. Routledge.

Barker, D. J. P. (2007). The origins of the developmental origins theory. *Journal of Internal Medicine, 261*(5), 412–417. https://doi.org/10.1111/j.1365-2796.2007.01809.x

Bassett, D. S., & Bullmore, E. T. (2016). Small-world brain networks revisited. *The Neuroscientist: A Review Journal Bringing Neurobiology, Neurology and Psychiatry, 23*(5), 499–516. https://doi.org/10.1177/1073858416667720

Bayraktar, G., Yuanxiang, P., Confettura, A. D., Gomes, G. M., Raza, S. A., Stork, O., Tajima, S., Suetake, I., Karpova, A., Yildirim, F., & Kreutz, M. R. (2020). Synaptic control of DNA methylation involves activity-dependent degradation of DNMT3A1 in the nucleus. *Neuropsychopharmacology, 45*(12), Article 12. https://doi.org/10.1038/s41386-020-0780-2

Bentivoglio, M., Cotrufo, T., Ferrari, S., Tesoriero, C., Mariotto, S., Bertini, G., Berzero, A., & Mazzarello, P. (2019). The original histological slides of Camillo Golgi and his discoveries on neuronal structure. *Frontiers in Neuroanatomy, 13*, 3. https://www.frontiersin.org/articles/10.3389/fnana.2019.00003

Bianco-Miotto, T., Craig, J. M., Gasser, Y. P., van Dijk, S. J., & Ozanne, S. E. (2017). Epigenetics and DOHaD: From basics to birth and beyond. *Journal of Developmental Origins of Health and Disease, 8*(5), 513–519. https://doi.org/10.1017/S2040174417000733

Binder, M. D., Hirokawa, N., & Windhorst, U. (Eds.). (2009). Neuron. In *Encyclopedia of neuroscience* (p. 2751). Springer. https://doi.org/10.1007/978-3-540-29678-2_3902

Bird, A. (2007). Perceptions of epigenetics. *Nature, 447*(7143), Article 7143. https://doi.org/10.1038/nature05913

Bock, J., Wainstock, T., Braun, K., & Segal, M. (2015). Stress in utero: Prenatal programming of brain plasticity and cognition. *Biological Psychiatry, 78*(5), 315–326. https://doi.org/10.1016/j.biopsych.2015.02.036

Bockmühl, Y., Patchev, A. V., Madejska, A., Hoffmann, A., Sousa, J. C., Sousa, N., Holsboer, F., Almeida, O. F. X., & Spengler, D. (2015). Methylation at the CpG island shore region upregulates Nr3c1 promoter activity after early-life stress. *Epigenetics, 10*(3), 247–257. https://doi.org/10.1080/15592294.2015.1017199

Bota, M., & Swanson, L. W. (2007). The neuron classification problem. *Brain Research Reviews, 56*(1), 79–88. https://doi.org/10.1016/j.brainresrev.2007.05.005

Bower, J. M., & Beeman, D. (Eds.). (1998). *The book of GENESIS* (2nd ed.). Springer. https://doi.org/10.1007/978-1-4612-1634-6

Bullock, T. H., Bennett, M. V. L., Johnston, D., Josephson, R., Marder, E., & Fields, R. D. (2005). The neuron doctrine, redux. *Science, 310*(5749), 791. https://doi.org/10.1126/science.1114394

Campbell, K. (2003). Signaling to and from radial glia. *Glia, 43*(1), 44–46. https://doi.org/10.1002/glia.10247

Chédin, F. (2011). The DNMT3 family of mammalian de novo DNA methyltransferases. *Progress in Molecular Biology and Translational Science, 101*, 255–285. https://doi.org/10.1016/B978-0-12-387685-0.00007-X

Chivet, M., Hemming, F., Pernet-Gallay, K., Fraboulet, S., & Sadoul, R. (2012). Emerging role of neuronal exosomes in the central nervous system. *Frontiers in Physiology, 3*, 145. https://doi.org/10.3389/fphys.2012.00145

Cimino, G. (1999). Reticular theory versus neuron theory in the work of Camillo Golgi. *Physis; Rivista Internazionale Di Storia Della Scienza, 36*(2), 431–472.

Cowan, W., & Kandel, E. (2001). Prospects for neurology and psychiatry. *JAMA: The Journal of the American Medical Association, 285*, 594–600. https://doi.org/10.1001/jama.285.5.594

Dahmen, D., Layer, M., Deutz, L., Dąbrowska, P. A., Voges, N., von Papen, M., Brochier, T., Riehle, A., Diesmann, M., Grün, S., & Helias, M. (2022). Global organization of neuronal activity only requires unstructured local connectivity. *eLife, 11*, e68422. https://doi.org/10.7554/eLife.68422

Darmanis, S., Sloan, S. A., Zhang, Y., Enge, M., Caneda, C., Shuer, L. M., Hayden Gephart, M. G., Barres, B. A., & Quake, S. R. (2015). A survey of human brain transcriptome diversity at the single cell level. *Proceedings of the National Academy of Sciences, 112*(23), 7285–7290. https://doi.org/10.1073/pnas.1507125112

Day, J. J., & Sweatt, J. D. (2010). DNA methylation and memory formation. *Nature Neuroscience, 13*(11), 1319–1323. https://doi.org/10.1038/nn.2666

Day, J. J., & Sweatt, J. D. (2011). Cognitive neuroepigenetics: A role for epigenetic mechanisms in learning and memory. *Neurobiology of Learning and Memory, 96*(1), 2–12. https://doi.org/10.1016/j.nlm.2010.12.008

De Backer, J.-F., & Grunwald Kadow, I. C. (2022). A role for glia in cellular and systemic metabolism: Insights from the fly. *Current Opinion in Insect Science, 53*, 100947. https://doi.org/10.1016/j.cois.2022.100947

Deichmann, U. (2016). Epigenetics: The origins and evolution of a fashionable topic. *Developmental Biology, 416*(1), 249–254. https://doi.org/10.1016/j.ydbio.2016.06.005

Dias, B. G., & Ressler, K. J. (2014). Parental olfactory experience influences behavior and neural structure in subsequent generations. *Nature Neuroscience, 17*(1), Article 1. https://doi.org/10.1038/nn.3594

Doi, S., & Kumagai, S. (2005). Generation of very slow neuronal rhythms and chaos near the Hopf bifurcation in single neuron models. *Journal of Computational Neuroscience, 19*(3), 325–356. https://doi.org/10.1007/s10827-005-2895-1

Elsayed, M., & Magistretti, P. J. (2015). A new outlook on mental illnesses: Glial involvement beyond the glue. *Frontiers in Cellular Neuroscience, 9*, 468. https://www.frontiersin.org/articles/10.3389/fncel.2015.00468

Emery, E. C., Luiz, A. P., Sikandar, S., Magnúsdóttir, R., Dong, X., & Wood, J. N. (2016). In vivo characterization of distinct modality-specific subsets of somatosensory neurons using GCaMP. *Science Advances, 2*(11), e1600990. https://doi.org/10.1126/sciadv.1600990

Enander, J. M. D., & Jörntell, H. (2019). Somatosensory cortical neurons decode tactile input patterns and location from both dominant and non-dominant digits. *Cell Reports, 26*(13), 3551–3560.e4. https://doi.org/10.1016/j.celrep.2019.02.099

Faber, D. S., & Pereda, A. E. (2018). Two forms of electrical transmission between neurons. *Frontiers in Molecular Neuroscience, 11*, 427. https://doi.org/10.3389/fnmol.2018.00427

Feng, J., Chang, H., Li, E., & Fan, G. (2005). Dynamic expression of de novo DNA methyltransferases Dnmt3a and Dnmt3b in the central nervous system. *Journal of Neuroscience Research, 79*(6), 734–746. https://doi.org/10.1002/jnr.20404

Fiedler, K., Kliegl, R., Lindenberger, U., Mausfeld, R., Mummendey, A., & Prinz, W. (2005). Psychologie im 21. Jahrhundert: Führende deutsche Psychologen über Lage und Zukunft ihres Fachs und die Rolle der psychologischen Grundlagenforschung. *Gehirn & Geist, 7–8*, 56–60.

Fields, R. D., & Stevens-Graham, B. (2002). New insights into neuron-glia communication. *Science (New York, N.Y.), 298*(5593), 556–562. https://doi.org/10.1126/science.298.5593.556

Fields, R. D., Woo, D. H., & Basser, P. J. (2015). Glial regulation of the neuronal connectome through local and long-distant communication. *Neuron, 86*(2), 374–386. https://doi.org/10.1016/j.neuron.2015.01.014

Fodstad, H. (2001). The neuron theory. *Stereotactic and Functional Neurosurgery, 77*(1–4), 20–24. https://doi.org/10.1159/000064596

Fraga, M. F., Ballestar, E., Paz, M. F., Ropero, S., Setien, F., Ballestar, M. L., et al. (2005). Epigenetic differences arise during the lifetime of monozygotic twins. *Proceedings of the National Academy of Sciences of the United States of America, 102*(30), 10604. https://doi.org/10.1073/pnas.0500398102

Franklin, T. B., & Mansuy, I. M. (2010). Epigenetic inheritance in mammals. *Neurobiology of Disease, 39*(1), 61–65. https://doi.org/10.1016/j.nbd.2009.11.012

Fries, P. (2015). Rhythms for cognition: Communication through coherence. *Neuron, 88*(1), 220–235. https://doi.org/10.1016/j.neuron.2015.09.034

Friston, K. J., Stephan, K. E., Montague, R., & Dolan, R. J. (2014). Computational psychiatry: The brain as a phantastic organ. *The Lancet Psychiatry, 1*(2), 148–158. https://doi.org/10.1016/S2215-0366(14)70275-5

Frühbeis, C., Fröhlich, D., Kuo, W. P., & Krämer-Albers, E.-M. (2013). Extracellular vesicles as mediators of neuron-glia communication. *Frontiers in Cellular Neuroscience, 7*, 182. https://doi.org/10.3389/fncel.2013.00182

Fuchs, T. (2018). *Ecology of the brain: The phenomenology and biology of the embodied mind* (First ed.). Oxford University Press.

Fuchs, T. (2020). The circularity of the embodied mind. *Frontiers in Psychology, 11*, 1707. https://doi.org/10.3389/fpsyg.2020.01707

Fuchs, E. C., Neitz, A., Pinna, R., Melzer, S., Caputi, A., & Monyer, H. (2016). Local and distant input controlling excitation in layer II of the medial entorhinal cortex. *Neuron, 89*(1), 194–208. https://doi.org/10.1016/j.neuron.2015.11.029

Gallese, V. (2013). Mirror neurons, embodied simulation and a second-person approach to mind-reading. *Cortex, 49*(10), 2954–2956. https://doi.org/10.1016/j.cortex.2013.09.008

Geyer, C. (Ed.). (2004). *Hirnforschung und Willensfreiheit: Zur Deutung der neuesten Experimente* (9. Auflage, Originalausgabe). Suhrkamp.

Golgi, C. (1885). *Sulla fina anatomia degli organi centrali del sistema nervosa.* Calderini e Figlio.

Gordon, E. (2003). Integrative neuroscience. *Neuropsychopharmacology, 28*(1), Article 1. https://doi.org/10.1038/sj.npp.1300136

Goto, K., Numata, M., Komura, J. I., Ono, T., Bestor, T. H., & Kondo, H. (1994). Expression of DNA methyltransferase gene in mature and immature neurons as well as proliferating cells in mice. *Differentiation, 56*(1–2), 39–44. https://doi.org/10.1046/j.1432-0436.1994.56120039.x

Goyal, D., Limesand, S. W., & Goyal, R. (2019). Epigenetic responses and the developmental origins of health and disease. *Journal of Endocrinology, 242*(1), T105–T119. https://doi.org/10.1530/JOE-19-0009

Gräff, J., & Mansuy, I. M. (2008). Epigenetic codes in cognition and behaviour. *Behavioural Brain Research, 192*(1), 70–87. https://doi.org/10.1016/j.bbr.2008.01.021

Guillery, R. W. (2005). Observations of synaptic structures: Origins of the neuron doctrine and its current status. *Philosophical Transactions of the Royal Society of London. Series B, Biological Sciences, 360*(1458), 1281–1307. https://doi.org/10.1098/rstb.2003.1459

Guillery, R. W. (2007). Relating the neuron doctrine to the cell theory. Should contemporary knowledge change our view of the neuron doctrine? *Brain Research Reviews, 55*(2), 411–421. https://doi.org/10.1016/j.brainresrev.2007.01.005

Haggard, P., & Libet, B. (2001). Conscious intention and brain activity. *Journal of Consciousness Studies, 8*(11), 47–63.

Hawrylycz, M. J., Lein, E. S., Guillozet-Bongaarts, A. L., Shen, E. H., Ng, L., Miller, J. A., et al. (2012). An anatomically comprehensive atlas of the adult human brain transcriptome. *Nature, 489*(7416), 391–399. https://doi.org/10.1038/nature11405

Heard, E., & Martienssen, R. A. (2014). Transgenerational epigenetic inheritance: Myths and mechanisms. *Cell, 157*(1), 95–109. https://doi.org/10.1016/j.cell.2014.02.045

Heim, C., & Binder, E. B. (2012). Current research trends in early life stress and depression. *Experimental Neurology, 233*(1), 102–111. https://doi.org/10.1016/j.expneurol.2011.10.032

Hines, M. L., & Carnevale, N. T. (2001). Neuron. *The Neuroscientist: A Review Journal Bringing Neurobiology, Neurology and Psychiatry, 7*(2), 123–135. https://doi.org/10.1177/107385840100700207

His, W. (2017). *Die Formentwickelung des menschlichen Vorderhirns: Vom Ende des ersten bis zum beginn des dritten Monats/Wilhelm His* (Nachdruck der Ausgabe von 1889). Hansebooks GmbH. http://nbn-resolving.de/urn:nbn:de:101:1-2018122721520047540280

Hoffman, D. J., Reynolds, R. M., & Hardy, D. B. (2017). Developmental origins of health and disease: Current knowledge and potential mechanisms. *Nutrition Reviews, 75*(12), 951–970. https://doi.org/10.1093/nutrit/nux053

Hormuzdi, S. G., Filippov, M. A., Mitropoulou, G., Monyer, H., & Bruzzone, R. (2004). Electrical synapses: A dynamic signaling system that shapes the activity of neuronal networks. *Biochimica et Biophysica Acta (BBA) – Biomembranes, 1662*(1), 113–137. https://doi.org/10.1016/j.bbamem.2003.10.023

Horsthemke, B. (2018). A critical view on transgenerational epigenetic inheritance in humans. *Nature Communications, 9*(1), Article 1. https://doi.org/10.1038/s41467-018-05445-5

Ibarra-Lecue, I., Haegens, S., & Harris, A. Z. (2022). Breaking down a rhythm: Dissecting the mechanisms underlying task-related neural oscillations. *Frontiers in Neural Circuits, 16*, 846905. https://www.frontiersin.org/articles/10.3389/fncir.2022.846905

Isles, A. R. (2015). Neural and behavioral epigenetics; what it is, and what is hype. *Genes, Brain, and Behavior, 14*(1), 64–72. https://doi.org/10.1111/gbb.12184

Isles, A. R., & Wilkinson, L. S. (2008). Epigenetics: What is it and why is it important to mental disease? *British Medical Bulletin, 85*(1), 35–45. https://doi.org/10.1093/bmb/ldn004

Jablonka, E., & Lamb, M. J. (2005). *Evolution in four dimensions: Genetic, epigenetic, behavioral, and symbolic variation in the history of life* (pp. x, 462). MIT Press.

Jones, E. G. (1994). The neuron doctrine 1891. *Journal of the History of the Neurosciences, 3*(1), 3–20. https://doi.org/10.1080/09647049409525584

Kahali, S., Raichle, M. E., & Yablonskiy, D. A. (2021). The role of the human brain neuron–glia–synapse composition in forming resting-state functional connectivity networks. *Brain Sciences, 11*(12), 1565. https://doi.org/10.3390/brainsci11121565

Kandel, E. R. (2000). *The molecular biology of memory storage: A dialogue between genes and synapses. Nobel lecture.* https://www.nobelprize.org/uploads/2018/06/kandel-lecture.pdf

Kato, Y., Kaneda, M., Hata, K., Kumaki, K., Hisano, M., Kohara, Y., Okano, M., Li, E., Nozaki, M., & Sasaki, H. (2007). Role of the Dnmt3 family in de novo methylation of imprinted and repetitive sequences during male germ cell development in the mouse. *Human Molecular Genetics, 16*(19), 2272–2280. https://doi.org/10.1093/hmg/ddm179

Kenet, T., Arieli, A., Tsodyks, M., & Grinvald, A. (2006). Are single cortical neurons soloists or are they obedient members of a huge orchestra? In J. L. van Hemmen & T. J. Sejnowski (Eds.), *23 Problems in systems neuroscience* (pp. 160–181). Oxford University Press. https://doi.org/10.1093/acprof:oso/9780195148220.003.0009

Koch, C. (1996). A neuronal correlate of consciousness? *Current Biology: CB, 6*(5), 492. https://doi.org/10.1016/s0960-9822(02)00519-5

Koch, C. (2019). *The feeling of life itself: Why consciousness is widespread but can't be computed.* MIT Press.

Kohler, E., Keysers, C., Umiltà, M. A., Fogassi, L., Gallese, V., & Rizzolatti, G. (2002). Hearing sounds, understanding actions: Action representation in mirror neurons. *Science (New York, N.Y.), 297*(5582), 846–848. https://doi.org/10.1126/science.1070311

Kolber, A. J. (2016). Free will as a matter of law. In D. Patterson & M. S. Pardo (Eds.), *Philosophical foundations of law and neuroscience* (pp. 9–28). Oxford University Press. https://doi.org/10.1093/acprof:oso/9780198743095.003.0002

Labonté, B., Suderman, M., Maussion, G., Lopez, J. P., Navarro-Sánchez, L., Yerko, V., Mechawar, N., Szyf, M., Meaney, M. J., & Turecki, G. (2013). Genome-wide methylation changes in the brains of suicide completers. *The American Journal of Psychiatry, 170*(5), 511–520. https://doi.org/10.1176/appi.ajp.2012.12050627

Levitan, I. B., & Kaczmarek, L. K. (2015). *The neuron: Cell and molecular biology* (Fourth ed.). Oxford University Press.

Libet, B. W. (1999). Do we have free will? *Journal of Consciousness Studies, 6*(8–9), 47–57.

Libet, B. W., Gleason, C. A., Wright, E. W., & Pearl, D. K. (1983). Time of conscious intention to act in relation to onset of cerebral activity (readiness-potential). The unconscious initiation of a freely voluntary act. *Brain: A Journal of Neurology, 106*(Pt 3), 623–642. https://doi.org/10.1093/brain/106.3.623

Lux, V. (2013). With Gottlieb beyond Gottlieb: The role of epigenetics in psychobiological development. *International Journal of Developmental Science, 7*(2), 69–78. https://doi.org/10.3233/DEV-1300073

Lux, V., & Richter, J. (Eds.). (2014). *Kulturen der Epigenetik: Vererbt, codiert, übertragen.* De Gruyter.

Maiers, W. (2009). Conceptual confusions in understanding human action and experience. In T. Teo, P. Stenner, A. Rutherford, E. Park, & C. Baerveldt (Eds.), *Varieties of theoretical psychology: International philosophical and practical concerns* (pp. 101–112). Captus University Publications.

Martin, R. (2019). *Neuroscience methods: A guide for advanced students.* CRC Press. https://doi.org/10.1201/9780367810665

McGowan, P. O., Sasaki, A., D'Alessio, A. C., Dymov, S., Labonté, B., Szyf, M., Turecki, G., & Meaney, M. J. (2009). Epigenetic regulation of the glucocorticoid receptor in human brain associates with childhood abuse. *Nature Neuroscience, 12*(3), 342–348. https://doi.org/10.1038/nn.2270

McKenna, T. M., McMullen, T. A., & Shlesinger, M. F. (1994). The brain as a dynamic physical system. *Neuroscience, 60*(3), 587–605. https://doi.org/10.1016/0306-4522(94)90489-8

McKenzie, A. T., Wang, M., Hauberg, M. E., Fullard, J. F., Kozlenkov, A., Keenan, A., Hurd, Y. L., Dracheva, S., Casaccia, P., Roussos, P., & Zhang, B. (2018). Brain cell type specific gene expression and co-expression network architectures. *Scientific Reports, 8*(1), Article 1. https://doi.org/10.1038/s41598-018-27293-5

Miller, E. K., & Cohen, J. D. (2001). An integrative theory of prefrontal cortex function. *Annual Review of Neuroscience, 24*, 167–202. https://doi.org/10.1146/annurev.neuro.24.1.167

Millman, D. J., Ocker, G. K., Caldejon, S., Kato, I., Larkin, J. D., Lee, E. K., Luviano, J., Nayan, C., Nguyen, T. V., North, K., Seid, S., White, C., Lecoq, J., Reid, C., Buice, M. A., & de Vries, S. E. (2020). VIP interneurons in mouse primary visual cortex selectively enhance responses to weak but specific stimuli. *eLife, 9*, e55130. https://doi.org/10.7554/eLife.55130

Moroz, L. L. (2018). NeuroSystematics and periodic system of neurons: Model vs reference species at single-cell resolution. *ACS Chemical Neuroscience, 9*(8), 1884–1903. https://doi.org/10.1021/acschemneuro.8b00100

Moroz, L. L. (2021). Multiple origins of neurons from secretory cells. *Frontiers in Cell and Developmental Biology, 9*, 669087. https://www.frontiersin.org/articles/10.3389/fcell.2021.669087

Mott, D. D., & Dingledine, R. (2003). Interneuron diversity series: Interneuron research – challenges and strategies. *Trends in Neurosciences, 26*(9), 484–488. https://doi.org/10.1016/S0166-2236(03)00200-5

Mountcastle, V. B. (1957). Modality and topographic properties of single neurons of cat's somatic sensory cortex. *Journal of Neurophysiology, 20*(4), 408–434. https://doi.org/10.1152/jn.1957.20.4.408

Murgatroyd, C., Patchev, A. V., Wu, Y., Micale, V., Bockmühl, Y., Fischer, D., Holsboer, F., Wotjak, C. T., Almeida, O. F. X., & Spengler, D. (2009). Dynamic DNA methylation programs persistent adverse effects of early-life stress. *Nature Neuroscience, 12*(12), 1559–1566. https://doi.org/10.1038/nn.2436

Murphy, N. C., Ellis, G. F. R., & O'Connor, T. (Eds.). (2009). *Downward causation and the neurobiology of free will.* Springer Verlag.

Nagelhus, E. A., Amiry-Moghaddam, M., Bergersen, L. H., Bjaalie, J. G., Eriksson, J., Gundersen, V., Leergaard, T. B., Morth, J. P., Storm-Mathisen, J., Torp, R., Walhovd, K. B., & Tønjum, T. (2013). The glia doctrine: Addressing the role of glial cells in healthy brain ageing. *Mechanisms of Ageing and Development, 134*(10), 449–459. https://doi.org/10.1016/j.mad.2013.10.001

Naumova, O. Y., Lee, M., Rychkov, S. Y., Vlasova, N. V., & Grigorenko, E. L. (2013). Gene expression in the human brain: The current state of the study of specificity and spatio-temporal dynamics. *Child Development, 84*(1), 76–88. https://doi.org/10.1111/cdev.12014

Negi, S. K., & Guda, C. (2017). Global gene expression profiling of healthy human brain and its application in studying neurological disorders. *Scientific Reports, 7*(1), Article 1. https://doi.org/10.1038/s41598-017-00952-9

Norris, G. T., Smirnov, I., Filiano, A. J., Shadowen, H. M., Cody, K. R., Thompson, J. A., Harris, T. H., Gaultier, A., Overall, C. C., & Kipnis, J. (2018). Neuronal integrity and complement control synaptic material clearance by microglia after CNS injury. *The Journal of Experimental Medicine, 215*(7), 1789–1801. https://doi.org/10.1084/jem.20172244

Okano, M., Xie, S., & Li, E. (1998). Cloning and characterization of a family of novel mammalian DNA (cytosine-5) methyltransferases. *Nature Genetics, 19*(3), 219–220. https://doi.org/10.1038/890

Palma-Gudiel, H., Córdova-Palomera, A., Leza, J. C., & Fañanás, L. (2015). Glucocorticoid receptor gene (NR3C1) methylation processes as mediators of early adversity in stress-related disorders causality: A critical review. *Neuroscience and Biobehavioral Reviews, 55*, 520–535. https://doi.org/10.1016/j.neubiorev.2015.05.016

Pascual, M., Ibáñez, F., & Guerri, C. (2019). Exosomes as mediators of neuron-glia communication in neuroinflammation. *Neural Regeneration Research, 15*(5), 796–801. https://doi.org/10.4103/1673-5374.268893

Pauen, M. (2007). Self-determination free will, responsibility, and determinism. *Synthesis Philosophica, 22*, 455–475+510.

Penfield, W., & Boldrey, E. (1937). Somatic motor and sensory representation in the cerebral cortex of man as studied by electrical stimulation. *Brain: A Journal of Neurology, 60*, 389–443. https://doi.org/10.1093/brain/60.4.389

Penfield, W., & Rasmussen, T. (1950). *The cerebral cortex of man; a clinical study of localization of function* (pp. xv, 248). Macmillan.

Peng, H., Xie, P., Liu, L., Kuang, X., Wang, Y., Qu, L., et al. (2021). Morphological diversity of single neurons in molecularly defined cell types. *Nature, 598*(7879), 174–181. https://doi.org/10.1038/s41586-021-03941-1

Pfeifer, R., Bongard, J., & Grand, S. (2007). *How the body shapes the way we think: A new view of intelligence*. MIT Press.

Pigliucci, M., Müller, G., & Konrad Lorenz Institute for Evolution and Cognition Research (Eds.). (2010). *Evolution, the extended synthesis*. MIT Press.

Pockett, S., Banks, W. P., & Gallagher, S. (2006). *Does consciousness cause behavior?* MIT Press Ebsco Publishing [distributor]. https://search.ebscohost.com/login.aspx?direct=true&scope=si te&db=nlebk&db=nlabk&AN=156014

Prinz, W. (2008). Der Wille als Artefakt. In K.-S. Rehberg (Ed.), *Die Natur der Gesellschaft: Verhandlungen des 33. Kongresses der Deutschen Gesellschaft für Soziologie in Kassel 2006. Teilbd. 1 u. 2* (pp. 642–655). Campus Verl.

Ptak, C., & Petronis, A. (2010). Epigenetic approaches to psychiatric disorders. *Dialogues in Clinical Neuroscience, 12*(1), 25–35.

Raichle, M. E. (2009). A paradigm shift in functional brain imaging. *The Journal of Neuroscience, 29*(41), 12729–12734. https://doi.org/10.1523/JNEUROSCI.4366-09.2009

Rajendran, L., Bali, J., Barr, M. M., Court, F. A., Krämer-Albers, E.-M., Picou, F., Raposo, G., van der Vos, K. E., van Niel, G., Wang, J., & Breakefield, X. O. (2014). Emerging roles of extracellular vesicles in the nervous system. *Journal of Neuroscience, 34*(46), 15482–15489. https://doi.org/10.1523/JNEUROSCI.3258-14.2014

Rakic, P. (1981). Neuronal-glial interaction during brain development. *Trends in Neurosciences, 4*, 184–187. https://doi.org/10.1016/0166-2236(81)90060-6

Ramón y Cajal, S. (1893). *Manual de histología normal y técnica micrográfica*. Liberia de Pascual Aguilar. https://wellcomecollection.org/works/wegkqafc

Ramón y Cajal, S. (1900). *Studien über die Hirnrinde des Menschen*. Verlag von Johann Ambrosius Barth.

Ramón y Cajal, S. (1906). *The structure and connexions of neurons*. https://www.nobelprize.org/uploads/2018/06/cajal-lecture.pdf

Retzius, G. (1890). *Zur Kenntniss des Nervensystems der Crustaceen* (Vol. I). Samson & Wallin.

Riggs, A. D., Martienssen, R. A., & Russo, V. E. A. (1996). Introduction. In *Epigenetic mechanisms of gene regulation* (Vol. 32, pp. 1–4). Cold Spring Harbor Laboratory Press. https://cshmonographs.org/index.php/monographs/article/view/4519

Rizzolatti, G., & Craighero, L. (2004). The mirror-neuron system. *Annual Review of Neuroscience, 27*, 169–192. https://doi.org/10.1146/annurev.neuro.27.070203.144230

Rizzolatti, G., & Fogassi, L. (2014). The mirror mechanism: Recent findings and perspectives. *Philosophical Transactions of the Royal Society of London. Series B, Biological Sciences, 369*(1644), 20130420. https://doi.org/10.1098/rstb.2013.0420

Rizzolatti, G., & Sinigaglia, C. (2016). The mirror mechanism: A basic principle of brain function. *Nature Reviews. Neuroscience, 17*(12), 757–765. https://doi.org/10.1038/nrn.2016.135

Rock, R. B., Gekker, G., Hu, S., Sheng, W. S., Cheeran, M., Lokensgard, J. R., & Peterson, P. K. (2004). Role of microglia in central nervous system infections. *Clinical Microbiology Reviews, 17*(4), 942–964. https://doi.org/10.1128/CMR.17.4.942-964.2004

Roth, G. (2003). *Fühlen, Denken, Handeln: Wie das Gehirn unser Verhalten steuert* (1. Aufl., [Nachdr.]). Suhrkamp.

Schmidt, U., Holsboer, F., & Rein, T. (2011). Epigenetic aspects of posttraumatic stress disorder. *Disease Markers, 30*(2–3), 77–87. https://doi.org/10.3233/DMA-2011-0749

Schmidt-Wilcke, T., Fuchs, E., Funke, K., Vlachos, A., Muller-Dahlhaus, F., Puts, N. A. J., Harris, R. E., & Edden, R. A. E. (2018). GABA-from inhibition to cognition. *The Neuroscientist: A Review Journal Bringing Neurobiology, Neurology and Psychiatry, 24*(5), 501–515. https://doi.org/10.1177/1073858417734530

Schöner, G. (2020). The dynamics of neural populations capture the laws of the mind. *Topics in Cognitive Science, 12*(4), 1257–1271. https://doi.org/10.1111/tops.12453

Sepulcre, J., Liu, H., Talukdar, T., Martincorena, I., Yeo, B. T. T., & Buckner, R. (2010). The organization of local and distant functional connectivity in the human brain. *PLoS Computational Biology, 6*, e1000808. https://doi.org/10.1371/journal.pcbi.1000808

Sharpee, T. O. (2014). Toward functional classification of neuronal types. *Neuron, 83*(6), 1329–1334. https://doi.org/10.1016/j.neuron.2014.08.040

Sherrington, C. S. (1906). *The integrative action of the nervous system* (pp. xvi, 411). Yale University Press. https://doi.org/10.1037/13798-000

Sidiropoulou, K., Pissadaki, E. K., & Poirazi, P. (2006). Inside the brain of a neuron. *EMBO Reports, 7*(9), 886–892. https://doi.org/10.1038/sj.embor.7400789

Singer, W. (1999). Neuronal synchrony. *Neuron, 24*(1), 49–65. https://doi.org/10.1016/S0896-6273(00)80821-1

Singer, W. (2004). Verschaltungen legen uns fest: Wir sollten aufhören, von Freiheit zu sprechen. In C. Geyer (Ed.), *Hirnforschung Und Willensfreiheit* (pp. 30–65). Suhrkamp.

Singer, W. (2013). Cortical dynamics revisited. *Trends in Cognitive Sciences, 17*(12), 616–626. https://doi.org/10.1016/j.tics.2013.09.006

Sjöstedt, E., Zhong, W., Fagerberg, L., Karlsson, M., Mitsios, N., Adori, C., et al. (2020). An atlas of the protein-coding genes in the human, pig, and mouse brain. *Science, 367*(6482), eaay5947. https://doi.org/10.1126/science.aay5947

Soon, C., Brass, M., Heinze, H.-J., & Haynes, J.-D. (2008). Unconscious determinants of free decisions in the human brain. *Nature Neuroscience, 11*, 543–545. https://doi.org/10.1038/nn.2112

Sporns, O. (2013). Structure and function of complex brain networks. *Dialogues in Clinical Neuroscience, 15*(3), 247–262.

Sporns, O., Tononi, G., & Kötter, R. (2005). The human connectome: A structural description of the human brain. *PLoS Computational Biology, 1*(4), e42. https://doi.org/10.1371/journal.pcbi.0010042

Stankiewicz, A. M., Swiergiel, A. H., & Lisowski, P. (2013). Epigenetics of stress adaptations in the brain. *Brain Research Bulletin, 98*, 76–92. https://doi.org/10.1016/j.brainresbull.2013.07.003

Stevens, B. (2003). Glia: Much more than the neuron's side-kick. *Current Biology, 13*(12), R469–R472. https://doi.org/10.1016/S0960-9822(03)00404-4

Sultan, F. A., & Day, J. J. (2011). Epigenetic mechanisms in memory and synaptic function. *Epigenomics, 3*(2), 157–181. https://doi.org/10.2217/epi.11.6

Sussillo, D. (2014). Neural circuits as computational dynamical systems. *Current Opinion in Neurobiology, 25*, 156–163. https://doi.org/10.1016/j.conb.2014.01.008

Tal, Z., Geva, R., & Amedi, A. (2017). Positive and negative somatotopic BOLD responses in contralateral versus ipsilateral penfield homunculus. *Cerebral Cortex, 27*(2), 962–980. https://doi.org/10.1093/cercor/bhx024

ten Oever, S., & Sack, A. T. (2019). Interactions between rhythmic and feature predictions to create parallel time-content associations. *Frontiers in Neuroscience, 13*, 791. https://www.frontiersin.org/articles/10.3389/fnins.2019.00791

Tognini, P., Napoli, D., & Pizzorusso, T. (2015). Dynamic DNA methylation in the brain: A new epigenetic mark for experience-dependent plasticity. *Frontiers in Cellular Neuroscience, 9*, 331. https://www.frontiersin.org/articles/10.3389/fncel.2015.00331

Uhlén, M., Fagerberg, L., Hallström, B. M., Lindskog, C., Oksvold, P., Mardinoglu, A., et al. (2015). Tissue-based map of the human proteome. *Science, 347*(6220), 1260419. https://doi.org/10.1126/science.1260419

Vaaga, C. E., Borisovska, M., & Westbrook, G. L. (2014). Dual-transmitter neurons: Functional implications of co-release and co-transmission. *Current Opinion in Neurobiology, 29*, 25–32. https://doi.org/10.1016/j.conb.2014.04.010

Van Gebuchten, A. (1891). La structure des centres nerveux. La moiille epiniere et le cervelet. *Cellule, 7*, 79–122.

van Hemmen, J. L., & Sejnowski, T. J. (Eds.). (2006). *23 Problems in systems neuroscience* (pp. xvi, 514). Oxford University Press. https://doi.org/10.1093/acprof:oso/9780195148220.001.0001

Varela, F. J., Thompson, E., & Rosch, E. (2016). *The embodied mind: Cognitive science and human experience* (Revised ed.). MIT Press.

von Kölliker, A. (1890). *Zur feineren Anatomie des centralen Nervensystems*. Wilhelm Engelmann. https://wellcomecollection.org/works/ycqnc9cv

Wadhwa, P. D., Buss, C., Entringer, S., & Swanson, J. M. (2009). Developmental origins of health and disease: Brief history of the approach and current focus on epigenetic mechanisms. *Seminars in Reproductive Medicine, 27*(5), 358–368. https://doi.org/10.1055/s-0029-1237424

Wagner, H., & Gaese, B. (2006). Can we understand the action of brains in natural environments? In J. L. van Hemmen & T. J. Sejnowski (Eds.), *23 Problems in systems neuroscience* (pp. 22–43). Oxford University Press. https://doi.org/10.1093/acprof:oso/9780195148220.003.0002

Waldeyer, W. (1891). *Ueber einige neuere Forschungen im Gebiete der Anatomie des Centralnervensystems.* Verlag von Georg Thieme.

Waterland, R. A., & Jirtle, R. L. (2003). Transposable elements. *Molecular and Cellular Biology, 23*(15), 5293–5300. https://doi.org/10.1128/MCB.23.15.5293-5300.2003

Weaver, I. C. G., Cervoni, N., Champagne, F. A., D'Alessio, A. C., Sharma, S., Seckl, J. R., Dymov, S., Szyf, M., & Meaney, M. J. (2004). Epigenetic programming by maternal behavior. *Nature Neuroscience, 7*(8), 847–854. https://doi.org/10.1038/nn1276

Xiong, H., & Gendelman, H. E. (Eds.). (2014). *Current laboratory methods in neuroscience research.* Springer.

Xu, Y., Jia, Y., Ma, J., Hayat, T., & Alsaedi, A. (2018). Collective responses in electrical activities of neurons under field coupling. *Scientific Reports, 8*(1), Article 1. https://doi.org/10.1038/s41598-018-19858-1

Yamamuro, K., Kimoto, S., Rosen, K. M., Kishimoto, T., & Makinodan, M. (2015). Potential primary roles of glial cells in the mechanisms of psychiatric disorders. *Frontiers in Cellular Neuroscience, 9*, 154. https://doi.org/10.3389/fncel.2015.00154

Yehuda, R., & Lehrner, A. (2018). Intergenerational transmission of trauma effects: Putative role of epigenetic mechanisms. *World Psychiatry, 17*(3), 243–257. https://doi.org/10.1002/wps.20568

Yehuda, R., Daskalakis, N. P., Bierer, L. M., Bader, H. N., Klengel, T., Holsboer, F., & Binder, E. B. (2016). Holocaust exposure induced intergenerational effects on FKBP5 methylation. *Biological Psychiatry, 80*(5), 372–380. https://doi.org/10.1016/j.biopsych.2015.08.005

Yehuda, R., Lehrner, A., & Bierer, L. M. (2018). The public reception of putative epigenetic mechanisms in the transgenerational effects of trauma. *Environmental. Epigenetics, 4*(2), dvy018. https://doi.org/10.1093/eep/dvy018

Yuste, R., Hawrylycz, M., Aalling, N., Aguilar-Valles, A., Arendt, D., Armañanzas, R., et al. (2020). A community-based transcriptomics classification and nomenclature of neocortical cell types. *Nature Neuroscience, 23*(12), Article 12. https://doi.org/10.1038/s41593-020-0685-8

Zeharia, N., Hertz, U., Flash, T., & Amedi, A. (2012). Negative blood oxygenation level dependent homunculus and somatotopic information in primary motor cortex and supplementary motor area. *Proceedings of the National Academy of Sciences of the United States of America, 109*, 18565–18570. https://doi.org/10.1073/pnas.1119125109

Zeharia, N., Hertz, U., Flash, T., & Amedi, A. (2015). New whole-body sensory-motor gradients revealed using phase-locked analysis and verified using multivoxel pattern analysis and functional connectivity. *Journal of Neuroscience, 35*(7), 2845–2859. https://doi.org/10.1523/JNEUROSCI.4246-14.2015

Chapter 2
Levels of Plasticity

Plasticity, then, in the wide sense of the word, means the possession of a structure weak enough to yield to an influence, but strong enough not to yield all at once. Each relatively stable phase of equilibrium in such a structure is marked by what we may call a new set of habits. Organic matter, especially nervous tissue, seems endowed with a very extraordinary degree of plasticity of this sort. (James, 1890, p. 105)

Our brain is flexible and plastic across the lifespan. This and similar notions stand for what has been called a "revolution" in modern neuroscience, prompted by neuroscientific findings that showed a tremendous recovery and trainability of mental functions in the adult brain. Before, plasticity was thought to be a property of the young brain, with early childhood staging all the critical and sensitive periods of brain development. It is now known that the adult and even the old brain are also plastic, being able to change synaptic plasticity as part of memory formation and learning (Bailey et al., 1996; Kandel, 2000), reorganize neural connections and functional representations (Merzenich, 2013; Merzenich & DeCharms, 1996; Merzenich & Jenkins, 1993), and even develop new neurons (Prochiantz, 2012; Prochiantz & Di Nardo, 2015). Moreover, some forms of neural flexibility observed in the adult brain are not present in the infant brain but are themselves the result of neural development (Yin et al., 2020). These findings changed our understanding of brain plasticity. The brain appears to be in constant flux, and the assumption of localizable brain functions is fundamentally challenged. Instead, it is assumed that function sits in the neural connections—"the connectome" (Hagmann, 2005; Hagmann et al., 2010; Sporns et al., 2005)—and that these connections are built and rebuilt throughout the lifespan. In the old battle between localizationism and holism in theories of consciousness, holism seems to gain ground again.

A closer look reveals that besides all this "network talk," there is still some localizationism (Kaiser & Cromby, 2014), or, more precisely, atomism grounding the current vision of the brain. The underlying concept of plasticity is one of neuronal connections with the single neuron as basic unit and connection hub. This neuron-based notion of plasticity was first fully elaborated by Donald O. Hebb as a mere

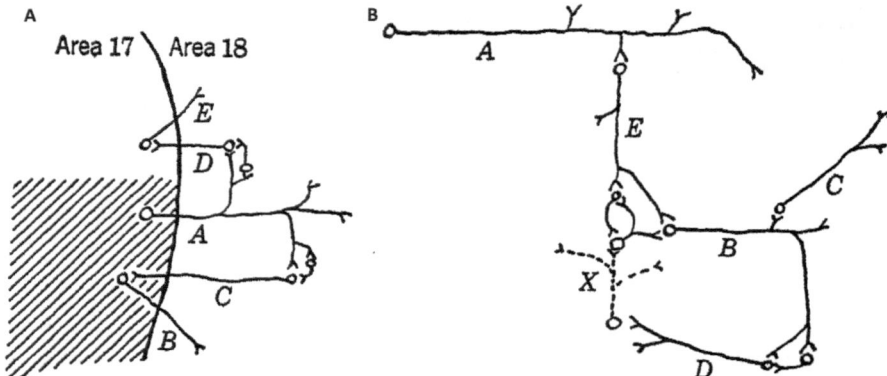

Fig. 2.1 Two drawings by Donald O. Hebb illustrating the formation of neuronal cell assemblies. (**a**) In the accompanying text, Hebb describes that due to the parallel excitation of neurons A and B, the synapses that connect to C are reinforced, and growth changes at the synapses will occur. As a consequence, neurons A and B will be coupled and no longer act independently from each other. In contrast, the connection between A and D will not undergo such a change, because other than A and B, E is outside the stimulated (hatched) area (Hebb, 1949, pp. 70–71). (**b**) Here, Hebb depicted the process of integration leading to coordinated activity of neurons of the same order stimulated by the same input (e.g., sensory neurons firing due to a particular visual stimulation). A, B C, D, and X represent possible direct and indirect connections between neurons in the specific brain area. In the diagram, A contributes to the firing of E, and B contributes to the firing of C and D. Hebb speculates that this is followed by growth changes at the synapses AE, BC, and BD. These changes then increase the probability of coordinated activity of these neuron pairs and, due to the connection of E and B, ultimately the overall neuron ensemble (Hebb, 1949, p. 72). (Reproduced with permission of Taylor and Francis Group LLC (Books) US through PLSclear)

theoretical assumption.[1] In his book *The Organization of Behavior* (1949), he outlined what later became known as Hebbian theory or rule:

> When an axon of cell A is near enough to excite a cell B and repeatedly or persistently takes part in firing it, some growth process or metabolic change takes place on one or both cells such that A's efficiency, as one of the cells firing B, is increased. (Hebb, 1949, p. 62)

Hebb further speculated that such a connection between neurons leads to the development or enlargement of "synaptic knobs" between A and B, which he assumed to be the physiological basis of their attuned activity (1949, p. 63). He also concluded that these dynamic adaptations to neural activity represent the neurological basis of learning: "In short, it is feasible to assume that synaptic knobs develop with neural activity and represent a lowered synaptic resistance. It is implied that the knobs appear in the course of learning" (1949, p. 66; see Fig. 2.1). Hebb described what is discussed today as synaptic plasticity, with one neuron inducing

---

[1] Although Hebb is often cited as having introduced this notion of plasticity, it seems more precise to understand his role as the one elaborating this notion and emphasizing its importance for understanding learning and the development of habits and behavioral patterns (see, for example, Berlucchi & Buchtel, 2009).

increased synaptic sensitivity to a stimulus onto another neuron via its own neural activity. Several experimental studies support the Hebbian theory (for review, see Bi & Poo, 2001). Most importantly, there is consensus that Hebbian plasticity underlies the development of long-term potentiation (LTP) and long-term depression (LTD), contributing to the neural basis of memory formation (Bailey & Kandel, 1995) and motor learning (Ziemann et al., 2006). In addition to Hebbian plasticity, anti-Hebbian plasticity has been observed in the brain. Anti-Hebbian plasticity is also based on synaptic connections, but the activity of the pre-synaptic neuron induces changes in the excitability or inhibition of the post-synaptic neuron while the post-synaptic neuron is at rest; thus, the neurons do not fire together (Kullmann & Lamsa, 2007). Hebbian and anti-Hebbian plasticity, as defined in this sense, also rely on different molecular mechanisms.[2] Their coexistence already complicates what is meant by plasticity at the level of synaptic connections.

Synaptic connections are not the only entities that exhibit plasticity in the brain and throughout an organism. Looking across the life sciences, we get the impression that plasticity is everywhere. Synapses exhibit plasticity in multiple forms (Citri & Malenka, 2008), but neurons also exhibit forms of non-synaptic plasticity (Mozzachiodi & Byrne, 2010; Zhang & Linden, 2003). Brain areas exhibit plasticity (Merzenich & DeCharms, 1996; Merzenich & Jenkins, 1993), and so do neural networks underlying cognitive functions (Brod et al., 2017; Guerra-Carrillo et al., 2014) as well as the whole phenotype (West-Eberhard, 2003, 2005). With the growing knowledge of epigenetic mechanisms, the notion of "epigenetic plasticity" is discussed as a way to describe changes at the epigenome level that influence gene expression and are related to cellular functions (Flavahan, 2020; Flavahan et al., 2017; Lazaris et al., 2020).

The different notions of plasticity at these different biological levels have in common that they emphasize a certain degree of changeability in response to external events detected by the organism via some stimulus or input (Berlucchi & Buchtel, 2009). What differs are the material underpinnings of plasticity as well as the input stimulating the change, which can be physical, molecular, electrophysiological, psychological, or even sociocultural in nature. Hence, the concept of plasticity answers to an immense set of biological phenomena and, on some occasions, has even become a substitute for other concepts, such as learning, development, or adaptation (Bateson & Gluckman, 2012). To understand and evaluate the notion of plasticity ascribed to the neuron and to neuroepigenetic mechanisms, we must

---

[2] Others have used the term "anti-Hebbian plasticity" to describe the inhibition of the postsynaptic neuron following the correlated activity with the presynaptic neuron, which Hebb did not account for in his model. Defined in this sense, anti-Hebbian plasticity has been interpreted as neural basis of LTD, with the difference between LTP and LTD induction stemming from the frequency and timing of the pre- and postsynaptic activity, while both rely on N-methyl-d-aspartate (NMDA) receptor activity (Choe, 2014; Kullmann & Lamsa, 2007). In contrast to this definition, however, I follow here the use of Kullmann and Lamsa (2007, p. 693), who argue that the term "anti-Hebbian plasticity" is more appropriate for cases where LTP or LTD are induced via calcium ion influx through glutamate receptors which occurs not while the postsynaptic neuron is firing but while it is at rest in its activation cycle.

disentangle the different concepts of plasticity used at these different levels. This is specifically relevant for our understanding of plasticity at the neuroepigenetic level because molecular epigenetic plasticity is interpreted as environmental sensitivity of epigenetic mechanisms and as such is interconnected with other forms of plasticity within the organism-environment system.

Plasticity as Developmental Resource

With the term "phenotypic plasticity," Mary Jane West-Eberhard (2003, p. 51) prominently characterized the ability of individual organisms to compensate for environmental, genetic, and physiological perturbations during their lifetime via phenotypic accommodation. According to this notion, an individual organism uses the range of phenotypic variability built into its developmental foundations to maintain a functional phenotype throughout its lifetime. In relation to the genetic level, for example, this includes phenomena which formerly have been discussed as "reaction norm"[3] (Klingenberg, 2019; Manuck, 2010). At the level of the immediate organismic environment, this includes niche construction or behavioral transmission within specific subpopulations (Keller & Ross, 1993; Stotz, 2017). In relation to the population level, one often-cited example of phenotypic plasticity is the change in sex rates due to water temperature in some fish or amphibian populations, which has recently been linked to epigenetic modifications (Budd et al., 2022; Flament, 2016; Todd et al., 2016). Overall, phenotypic plasticity represents the developmental variability of individual organisms or populations in their species-specific environment and may be based on all parts of their developmental system (Oyama, 2000). In this broad definition, as used by West-Eberhard and others, plasticity tends to replace development. Moreover, this notion of plasticity implicates a change of perspective in evolutionary biology by (re)emphasizing the role of ontogenetic development as evolutionary factor (as debated in full depth by Gilbert, 2001; Gilbert et al., 2015; Gilbert & Barresi, 2020; Odling-Smee et al., 2003; Oyama, 2000; Stotz, 2017).

   With this emphasis on ontogenetic variability as a developmental resource that enhances an organism's adaptability, the question arises as to why there should be limits to the changeability of the organism's biological constitution. West-Eberhard and others answer this in energetic terms: while reversible phenotypic plasticity would provide nearly unlimited phenotypic flexibility, constant phenotypic changes would involve large energetic costs for an individual (Piersma & Drent, 2003; West-Eberhard, 2003). In contrast, non-reversible developmental plasticity also produces well-adapted phenotypes but at a lower cost of phenotypic changes, as they must be realized only once during development. One example is the adaptation of metabolic

---

[3] The term "Reaktionsnorm" was originally coined in 1909 by Richard Woltereck (1913) in the context of phylogenetic development at the species level but has since also been used to describe ontogenetic variation.

systems to environmental conditions during critical periods of physiological set-point calibration in early organismic development. The setpoints influence the phenotype over the lifespan and provide an optimal adaptation in cases where the prenatal and early life environment is a reliable predictor of conditions experienced later in life (Gluckman et al., 2005; Hanson & Gluckman, 2014).[4] However, under changing environmental conditions, this early adaptation could become an unforeseen maladaptation (Hales & Barker, 1992).

This notion of adaptive physiological plasticity was one of the founding principles of the Developmental Origins of Health and Disease (DOHaD) framework developed by David Barker and others (Barker, 2007). This approach studies the effects of early life environmental conditions, including nutrition, hormonal levels, exposure to toxins, etc., on the development of complex diseases later in life from an epidemiological perspective. The DOHaD approach has specifically become known for the "mismatch hypothesis," which assumes that physiological differences due to differences in the early and later environmental conditions of an organism underlie disease risk later in life. For example, a physiological setpoint optimized for low-calorie intake due to food restrictions in the early environment increases the risk of diabetes, obesity, and the metabolic syndrome in an environment characterized by ample food availability. Several animal models have been used to study the relationship between early environmental adaptation and later disease development in accordance with this framework, most prominently for obesity and the metabolic syndrome, confirming that the physiological setpoints of the underlying metabolic functions play an important role in disease development (Hsu & Tain, 2021; McMullen & Mostyn, 2009). Still, the material underpinnings of this early life phenotypic plasticity as well as the sensitive period for its fine-tuning are not entirely uncovered. More recently, epigenetic programming effects, especially DNA methylation, have been discussed as the molecular underpinnings of these phenomena at the single-cell level (Bianco-Miotto et al., 2017; Goyal et al., 2019; Yamada & Chong, 2017). These are interpreted as environmental sensitive switch points for the developmental pathways of the different physiological systems, indicating that plasticity at the physiological or behavioral level is assumed to be reflected at the molecular and, particularly, the epigenetic level.

Experience-Dependent Plasticity of Neural Networks

In contrast to this broad notion of phenotypic plasticity, the notion of plasticity used to describe experience-dependent changes in psychological functions and related neural networks directly builds on the Hebbian notion of synaptic plasticity at the single neuron level. One example is the work of psychologist Silvia A. Bunge, who,

---

[4] Some of the notions in this paragraph have been inspired by discussions and scholarly exchanges with Barbara Tzschentke and E. Tobias Krause.

together with her group, studies neural plasticity related to cognitive abilities and reasoning from a developmental perspective. Bunge focuses on experience-dependent improvements in cognitive and memory capacities. In her studies, she identified socio-cultural settings that addressed learning and cognitive abilities, such as entering elementary school or participating in a cognitive skill training program, and then studied the improvement in general cognitive functions and the long-term effects beyond directly trained skills (Brod et al., 2017; Guerra-Carrillo et al., 2014; Mackey et al., 2013; Wendelken et al., 2016). Overall, the group found that learning generally improves learning (Bunge & Leib, 2020). Thus, if you are used to learn on a regular basis, you are better in learning in other settings as well as in related cognitive or memory tasks. They also found that learning improves cognitive abilities over the lifespan, although it cannot fully compensate for cognitive decline in old age (Prull et al., 2000). Furthermore, the ways of learning or what one learns impact cognitive abilities differently, depending on what is trained (Guerra-Carrillo et al., 2014).

The group also investigated how these learning experiences change the underlying neural networks (Brod et al., 2017; Mackey et al., 2013). For example, Brod et al. (2017) studied how schooling at the elementary level enhances the functional connectivity of brain regions previously associated with executive functions and self-control. In this approach, plasticity is situated at the neural-network level. It is interpreted as the mechanism underlying experience-dependent cognitive development and training effects and is realized by changing functional and structural connectivity patterns. As outlined in the founding editorial of the journal *Brain Connectivity*, within the connectome framework, these connectivity patterns are understood as dynamic changing connections and interactions between distinct units of the CNS, and these "units can be categorized into levels of micro (individual neurons), meso- (columns), or macro- (regions) scales" (Pawela & Biswal, 2011). In this framework, plasticity at the level of higher psychological functions, such as memory or cognition, is considered to reflect plasticity at the network or system level, and plasticity at the network level reflects plasticity at other scales, such as the region, column, or single-neuron level. This clearly echoes Hebb's idea of a neurological-based psychological learning theory, grounded in synaptic plasticity.

Similar assumptions underlie the neurodevelopmental concept of plasticity proposed by Merzenich et al. In a conceptual paper, Nahum, Lee, and Merzenich (2013) emphasize the connectivity and network perspective even stronger, stating that the brain does not comprise of distinct units but rather that brain function and the behavior resulting from it is located at the systems or network level:

> [O]ur expressive behaviors are a product of complex, multilevel recurrent networks [...]. It is important to understand that in these recursive recurrent networks, the operational levels contributing to the representation of any aspect of input or action in brain systems are inseparable; in other words, all explicit behaviors are a product of the system. (Nahum et al., 2013, pp. 142–243)

In this context, Nahum et al. also argue that plasticity is controlled in a "top-down" manner (Nahum et al., 2013, p. 143). Thus, lower-level plasticity (or change-ability) depends on the representation of these changes at higher levels of the system, asserting that lower-level plasticity is initiated and channeled by these higher-level representations. Still, according to their notion, the range of plasticity at the higher levels depends on the range of plasticity at lower network levels and the synaptic connections between single neurons: "Plasticity is primarily expressed by a change in connectional strength at the synapse level, achieved both by increasing the powers and the numbers of synapses specifically supporting a progressively improving behavior" (Nahum et al., 2013, p. 144). Accordingly, representations at higher system levels must be directly reflected in the activity of cortical neurons, and plasticity at higher levels strongly depends on the synaptic plasticity of these cortical neurons, which results from the neural input received from the network level in a recursive feedback cycle. Thus, although different levels of plasticity at different levels of the system are considered by the authors, it all comes back to the single neuron and its plasticity properties. Nahum et al. summarize this dependence of the higher level on the lower-level activity range as follows:

> Cortical neurons at all "higher" system levels are integrators operating with very short time constants. […] The greater the coordination of neurons in the lower levels of the network that feeds them, the greater their selective powers and selectivity, and the greater the power of that input to drive plastic remodeling at higher system levels. Moreover, at the "top" of our great brain systems, coordination of activity is a primary determinant of the power with which cortical networks can sustain the reverberant activities that are selective for behavioral targets or goals (i.e., working memory) […]. The strengths of these key plasticity-gating processes at the top are critically dependent upon the strengths of the coordinated inputs that feed them. (Nahum et al., 2013, p. 144)

Thus, although plasticity at the higher functional levels is understood by Nahum et al. to be reflected in synaptic properties at the single cell level—"increasing the powers and the numbers of synapses" (Nahum et al., 2013, p. 144)—it is not fully depending on a specific single neuron and its synaptic connections. Instead, it is modeled as a majority vote of single neurons that coordinate their activity via the strengthened neural connections at the synapse level.

Research in the field of sensory plasticity supports this recursive top-down notion, demonstrating a strong relationship between plasticity at the neural network level and neuronal and synaptic plasticity at the single-neuron level (Das, 1997). This research showed a tremendous degree of experience-dependent structural and functional changes in neural networks and neuronal connections, even in the adult brain. The findings indicate, for example, that the projection fields of the sensory cortices change, expand, or diminish as a direct effect of their current and former use (see, for example, Bengoetxea et al., 2012; Das, 1997; Guic et al., 2008; Ribic, 2020). While the effect of the sensory input is found to be strongest during critical and sensitive developmental periods of the specific sensory system, during which the cortical neurons are more susceptible to structural and functional alterations, they continue to occur throughout the life course (Elbert et al., 1995; Espinosa & Stryker, 2012; Greenough et al., 1987; Zhao et al., 2022). Studies conducted within the whisker-to-barrel system in rodent models have been pioneering in connecting functional changes, neural connectivity, and the single neuron and synapse level (Campelo et al., 2020; Erzurumlu & Gaspar, 2020). In humans, studies with professional musicians showed that experience and learning over the lifetime induce substantial changes in neural networks and neural activity patterns related to auditory perception (Koelsch et al., 1999; Rüsseler et al., 2001). In addition to the sensory system, experience-dependent structural changes have also been reported in brain areas related to memory processes. Here, a seminal study with London taxi drivers reported larger posterior hippocampi in the taxi drivers compared to the control subjects, and this difference was related to the time spent as taxi driver (Maguire et al., 2000).

In this context, professional musicians are debated as prime models for the study of experience-dependent neural plasticity in humans (Münte et al., 2002). Music not only represents a complex stimulus, but professional music production is also a complex behavioral task that affords the precise integration of sensory, motor, and executive functions only reached after years of intensive training. This allows to compare interaction and interconnectivity of neural network changes across different functional levels and modalities, and specifically across sensory, motor, and cognitive functions. Although limited to what is traceable with neural imaging methods, and thus restricted to the neural network and connectivity level, cross-sectional and longitudinal studies in this specific group address the full bandwidth as well as the limits of human neuroplasticity, including short-term changes related to learning a specific musical play, long-term changes related to years of practice, and aging effects. By accounting for the family environment and the onset of the musical training, also the role of critical and sensible periods of neurodevelopment in lifelong neuroplasticity can be studied. In addition, maladaptive neuroplasticity represented by training-induced changes in neural networks that are related to functional failure, such as the "musicians cramp," is also successfully studied in this group (see Münte et al., 2002 for review). Similarly, professional athletes provide a model for plasticity processes related to motor functions at different timeframes and neurophysiological levels, and their interrelationship with cognitive functions, mental health, and functional failure (Krüger & Lux, 2023; Seidel-Marzi & Ragert, 2020).

At first sight, the emerging evidence from this type of research investigating training or learning-based plasticity of sensory and motor networks as well as other brain structures stands in contrast with our knowledge about constrained developmental windows of, for example, the development of the visual system (Hooks & Chen, 2007; Hubel & Wiesel, 1963), the auditory system (Knudsen et al., 1984), and the olfactory system (Franks & Isaacson, 2005). Studies specifically investigating the difference between developmental and adult plasticity show that passive exposure to relevant sensory input is sufficient for changes in early life, while adult plasticity depends on higher-order attentional mechanisms and top-down regulation from higher neural networks (for an extensive review, see Ribic, 2020). This finding is consistent with the effects observed in professional musicians, athletes, and the taxi drivers. When trying to integrate these findings at a theoretical level, we come back to the neuron and its synaptic connections. Following the Hebbian notion, their degree of changeability is considered the ultimate basis for neuroplasticity feeding into the functionally specified higher-level networks. However, this raises the question of how plasticity at the level of coordinated neuron populations relates to the plasticity of a single neuron within this population. The single neuron must exhibit properties that integrate both plasticity, with regard to its excitability and synaptic connections, and non-plasticity, with regard to its, at least temporal, functional specification. So, what keeps the neuron changing and what keeps it unchanged?

Plasticity at the Single Neuron Level: Changing Excitability with Fixed Function

As the brain's plasticity is assumed to originate to a great part from the basic properties of the neuron, neurons themselves are depicted as "highly flexible" (Borrelli et al., 2008, p. 961) in their morphology, connections, and excitability. Among the diverse forms of plasticity exhibited by neurons, synaptic plasticity is the most researched. In addition, as discussed above, it is understood to contribute directly to plasticity at higher levels, such as the network level, cognition, or learning, as these forms of plasticity are ultimately considered grounded in synaptic plasticity.

First hints on the molecular basis of synaptic plasticity were originally discovered by Eric R. Kandel and his colleagues when experimenting with the seasnail *Aplysia*. In his Nobel Lecture, Kandel stressed that, at some point, they had reduced their research focus to the question how learning could take place in an invariant, stable neuronal network as the one underlying *Aplysia*'s gill-withdrawal reflex (Kandel, 2000; see Fig. 2.2). This reductionist approach to learning and memory focused on changes in the excitability of neurons and the molecular mechanisms involved in these changes at the synapses. They found that the underlying

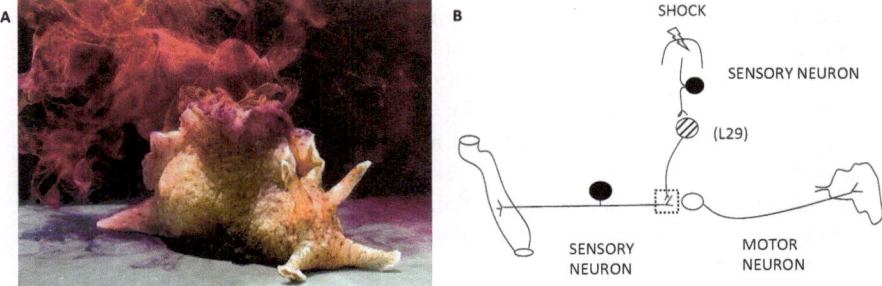

Fig. 2.2 The gill withdrawal reflex of *Aplysia californica*. (**a**) *Aplysia californica* emitting ink cloud. By Genny Anderson, CC BY-SA 4.0, http://marinebio.net/marinescience/03ecology/tptre. htm. (**b**) The neural connectivity of the *Aplysia* gill withdrawal reflex used for studying the process of sensitization. (Adapted from Seán Commins, 2018, p. 57 / Fig. 5.8. Reproduced with permission of Cambridge University Press through PLSclear)

mechanisms involved reversible changes in the synaptic connections between the involved neurons, and they were able to describe some of the contributing molecular pathways (see Box 2.1). Today, we know that synaptic plasticity is influenced by various alterations at the synapse level and includes structural and functional changes. This starts with the mere number of synapses and synaptic connections exhibited by a neuron and continues with modifications in the sensitivity and responsivity of these synapses. In addition to morphological diversity, variations can result from the pre-synaptic availability and the amount of released neurotransmitters, regulated, for example, via calcium influx but also genetic variation, as well as the number of post-synaptic receptors or ion channels, among other mechanisms (for an overview of different mechanisms participating in short- and long-term synaptic plasticity, see Citri & Malenka, 2008).

Box 2.1 Molecular Mechanisms in Synaptic Plasticity

Using *Aplysia*'s gill-withdrawal reflex as a model, Kandel et al. studied how acquired behavioral changes are reflected in the strength of synaptic connections between individual neurons. With its large neurons and a developmental program that ensured invariant connections between them, *Aplysia*'s neural network provided an optimal experimental system to study the underlying electrophysiological and molecular mechanisms. Due to the stability of the network, the observed changes could be assigned directly to alterations in the strength and effectiveness of the synaptic connections.

In their sensitization experiments, Kandel et al. trained the gill-withdrawal reflex by pairing a mild tactile stimulus applied to the snail's siphon with electroshocks applied to its tail. This increased the excitability at the synapses that connect the sensory neurons and the motor neurons in the reflex arc. They further investigated the underlying molecular changes. After a series of experiments, they found that serotonin is released by neighboring interneurons and acts on a transmembrane receptor of the sensory neuron. Subsequently, adenylyl cyclase (AC) is activated and ATP is converted to the intracellular second messenger adenosine $3',5'$-cyclic monophosphate (cAMP), which then activates cyclic cAMP-dependent protein kinase A (PKA). PKA phosphorylates K^+ channels and initiates additional modifications of the exocytotic machinery of the synapse, which, for example, enables greater Ca^{2+} influx into the presynaptic terminal with each action potential. Overall, this increases transmitter availability and release, resulting in increased excitability and a broadening of the action potential. Kandel et al. interpreted these molecular changes observed in the context of the short-term behavioral changes as basic form of short-term memory.

They further observed that when the sensitization treatment was continued for more than 1 day and electroshocks were repeatedly applied, PKA levels increased dramatically at some point, up to a level at which PKA is abundant, and together with mitogen-activated protein kinase (MAPK), which is recruited during the process, translocates to the nucleus. In the nucleus, they

(continued)

Box 2.1 (continued)

activate a transcriptional cascade involving cAMP response element-binding protein (CREB), which stabilizes persistent PKA activity and phosphorylation of its substrate proteins and initiates the synthesis of proteins that promote the growth of new synaptic connections. This then stabilizes the increased excitability of the synapse, even after the sensitization treatment is stopped. Accordingly, Kandel et al. interpreted this long-term sensitization as a basic form of long-term memory. They conclude that this form of synaptic plasticity provides a fundamental mechanism for information storage in the nervous system directly built into the molecular architecture of the chemical synapses.

For a detailed account of the individual experimental steps, see Kandel (2000).

Still, synaptic plasticity seems to explain only a portion of the observed changes in excitability and activation readiness of neurons. In recent years, various forms of non-synaptic plasticity were investigated in this regard. The term non-synaptic plasticity summarizes mechanisms that change the intrinsic excitability of a neuron, but do not originate from the synapse level (Mozzachiodi & Byrne, 2010). This includes changes in post-synaptic potentials due to changes in the number of ion channels in the axon, dendrites, or the soma of a neuron. These have been shown, for example, to influence spike generation (Hansel et al., 2001), contribute to long-term associative memory formation (Kemenes et al., 2006), and interact with synaptic plasticity in memory and learning (Mozzachiodi & Byrne, 2010). One important role of non-synaptic plasticity mechanisms is to facilitate homeostatic plasticity. Homeostatic plasticity provides some of the guardrails for the inner stability of a neuron necessary to maintain proper cell function (Turrigiano, 2017). The significance of homeostatic plasticity for neural function was first noticed in the context of cell culture studies. Neurons develop a high degree of excitability in vitro when their spontaneous activity is inhibited (Corner & Ramakers, 1992). In contrast, when stimulated over a longer period, their excitability decreases (Turrigiano et al., 1998). Accordingly, synaptic and homeostatic plasticity are partially contradictory mechanisms, and there are ongoing efforts to integrate both types of plasticity into a unified model of neuronal plasticity (Keck et al., 2017; Turrigiano, 2017; Turrigiano & Nelson, 2000).

In addition to synaptic and homeostatic plasticity, a neuron's excitability is also influenced by neuromodulation. Neighboring neurons can directly release neurotransmitters in the surrounding neural tissue, which then diffuse through the cell membrane of the non-releasing neurons and modulate neural function. This mechanism of neuromodulation can activate and deactivate neurons in milliseconds without changing synaptic weight (Scheler & Fellous, 2001). It represents a local yet fast-reacting system of neuronal activation, which adds to the variety of mechanisms influencing plasticity at the single neuron level.

Despite this tremendous flexibility in synaptic activity, excitability, and related homeostatic and anatomic features, neurons are depicted as relatively stable in their functional role once embedded within a specific neuronal network underlying a specific behavioral pattern. The high degree of specialization of neurons in the somatosensory and motor cortices somehow implied that a neuron's activity stays related to one behavioral pattern or sensory input. Thus, while the sensitivity to the signal is understood to adapt via the observed mechanisms of synaptic and non-synaptic plasticity to ensure signaling at optimized energetic costs, as in *Aplysia*'s gill withdrawal reflex, the specific function in which the neuron participates is thought to be fixed once the neuron is integrated in a specific signaling network. This view is in part the result of the reductionist approach taken by Kandel and his colleagues in choosing a stable neural network as model for their investigations (Kandel, 2000). Further support came from the finding that, in specific learning paradigms, changes in behavioral patterns were accompanied by the recruitment of alternative neural signaling pathways. The spontaneous recovery of extinguished associations, as seen in extinction learning experiments after certain time periods or in relation to context clues of the original association, could be well explained by this notion (Bouton et al., 2021; Dunsmoor et al., 2015). The neural basis of "learning" in this conditioning paradigm is thus understood as the formation and strengthening of new signaling pathways, which partly override established pathways weakened by the lack of the previous signal. The strengthening and weakening of these pathways are interpreted as the result of changes in synaptic sensitivity and, accordingly, have been shown to be enhanced or blocked via pharmacological manipulation of mechanisms regulating synaptic plasticity (for review, see Bouton et al., 2021). This interpretation is also in accordance with the notion that critical and sensitive periods for specific functions occur during early developmental periods, when most neurons are not yet fully integrated into functional specific tasks, and would therefore bear a stronger potential to newly connect and adapt to the required function. In contrast, at a later developmental stage, there would be fewer neurons available for new wiring, overall limiting neural plasticity (M. H. Johnson, 2011). Strong plasticity at the single-neuron and synaptic level and reduced plasticity at the neural network level outside critical periods are integrated in this view.

Plasticity as Functional Flexibility at the Single-Cell Level

One well-researched example of a highly specialized neural network related to a complex behavioral pattern, which has been in accordance with this view for a long time, is the song singing behavior of male zebra finches. Male zebra finches sing nearly the same song throughout their lives. They acquire it in early life during an intense period of song learning and reproduce it daily. The brain area and neuronal networks underlying this stable singing behavior are well described (see Fig. 2.3), and, according to this notion, the underlying neuronal activation was thought to be quite stable as well after being established in the sensitive period of song learning.

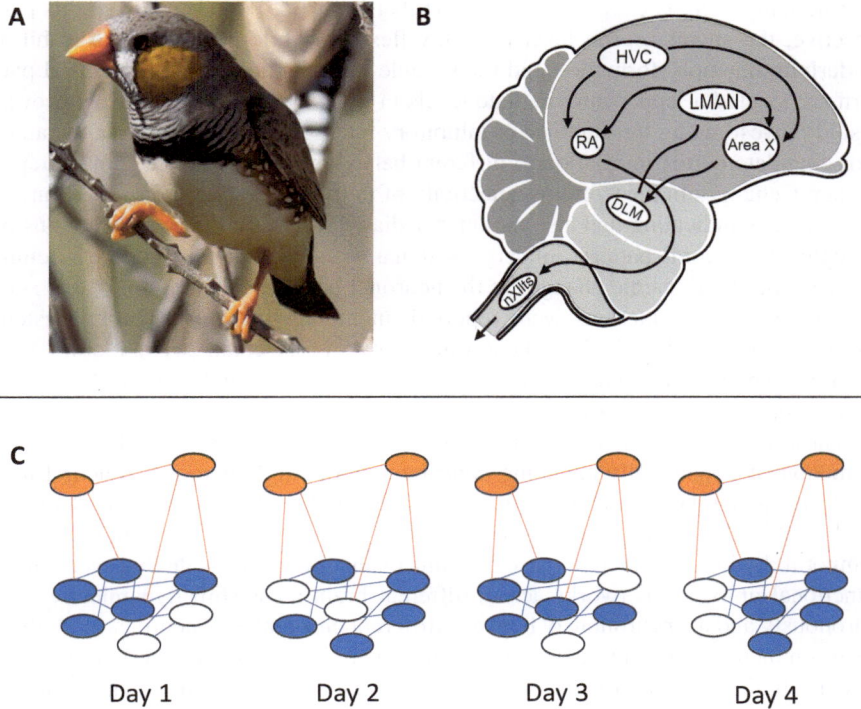

Fig. 2.3 Flexible participation of single neurons in the song-singing behavior of male zebra finch. (**a**) Male zebra finch. By Jim Bendon, CC BY-SA 2.0 Deed. (**b**) A schematic drawing of the neural network contributing to song learning and production in male zebra finch. LMAN: lateral magnocellular nucleus of the anterior nidopallium; RA: robust nucleus of the arcopallium; HVC: high vocal center, DLM: dorsolateral anterior thalamic nucleus, nXIIts: tracheosyringeal half of the hypoglossal nucleus. (Adapted from L. Shyamal based on Nottebohm, 2005, CC BY 2.5 Deed, grayscale.) (**c**) Flexible participation of single neurons in local neural networks in the HVC activated during song singing over several days. Blue: active neuron in the local network, white: nonactive neuron in the local network, orange: active interneuron. The graphic shows a highly simplified illustration of the mechanism reported by Liberti et al. (2016)

However, it turned out that things are different at the single-neuron level. The relationship seems to be more complicated, and inhibition seems to play an even more crucial role in the persistence of behavior than excitation. Using a multi-method approach to record the activity of inhibitory and excitatory neurons during song singing at the single-cell level over 30 days, Liberti et al. (2016) showed that the activity of single excitatory neurons associated with the stable reproduction of the song was only stable over the course of a single day but changed dramatically over the course of several days or weeks. In contrast, the involvement of inhibitory interneurons and the overall ensemble of neurons were relatively stable during the same period. The authors concluded that the identical reproduction of the song relied on the meso-spatial stability of the underlying neuronal network, with a highly flexible recruitment of single excitatory neurons (Liberti et al., 2016). Thus, from this perspective, the single neuron level is highly flexible, while the functional stability underlying memory formation and the reliable reproduction of the behavioral pattern are located at upper-intermediate levels in the network connections. However, a study investigating the stability of inhibitory interneuron functioning in the anterior cingulate cortex in relation to different behavioral tasks in mice showed rapid dynamic changes in the stimulus-selectivity of individual interneurons (C. Johnson et al., 2022), indicating that also the intermediate level of neural representations of complex behavioral patterns undergoes dynamic changes under certain circumstances. Similar dynamic changes in the neuronal basis of different motor and sensory systems and functions were reported, including the whisker-barrel system (Margolis et al., 2012), the lateral entorhinal cortex (Tsao et al., 2018), motor learning representation in the motor cortex (Peters et al., 2017), and memory formation in the hippocampus (Bladon et al., 2019).

These findings across different domains and brain areas indicate that the phenomenon of dynamically changing contributions of single neurons to neural networks is a general mechanism within the brain (Chambers & Rumpel, 2017; Clopath et al., 2017; Mau et al., 2020). Mau et al. (2020) emphasize the different scales and temporal rhythms involved in these dynamic changes and speculate that even whole functional units and networks across different brain areas shift dynamically synchronously and asynchronously between different ensembles of neurons. With their focus on memory, they propose to view this dynamic ensemble method of the brain as contributing to memory consolidation, catering to the observation that memories need maintenance and updating to be conserved (Mau et al., 2020). They further propose that the mechanisms of synaptic and non-synaptic plasticity contribute to this neural drift by providing the range of a neuron's readiness to be involved in a signaling pathway at a certain point in time.

The Role of Adult Neurogenesis in Brain Plasticity

Thus, at the single-neuron level, plasticity is present in the form of synaptic, non-synaptic, and homeostatic plasticity, as well as the degree of functional variability, as shown by the temporality of a neuron's participation in a specific signaling pathway and its ability to participate in different functional pathways. However, studies investigating the molecular basis of critical periods of neuron development, specifically neurogenesis, indicate that the number of available neurons is also dynamic over the lifespan of an individual, adding another form of plasticity at the single neuron level. The bio-philosopher Tobias Rees (2016) coined this neurogenesis-based plasticity "embryogenetic plasticity," emphasizing the developmental dimension in contrast to, for example, short-term changes in synaptic excitability. Alain Prochiantz and his colleagues, who study the role of homeoproteins in neurogenesis, are at the forefront of this new perspective. Their findings challenged the last notion of stability at the single-neuron level: that neurogenesis does not occur in the adult human brain. Until recently, it was widely accepted that neurogenesis is completed prenatally and that the number of neurons we are born with can only decrease during our lifetime. This notion grounds the tragedy of neurodegenerative diseases, such as Parkinson's and Alzheimer's, which are partly understood as the result of an increased rate of neurons dying in specific brain regions.

Prochiantz et al. found that homeoproteins, which participate in embryonic neurogenesis, can easily diffuse into the nuclei of adult neurons. They assumed that this presents a protein-based cell communication mechanism that is still able to act in the adult brain (Prochiantz & Di Nardo, 2015). They also showed in vitro that homeoproteins initiate the newly differentiation of neuronal precursor cells into neurons, and they provided first evidence that some of these homeoproteins, such as orthodenticle homeobox 2 (Otx2) and engrailed 1/2 (En1/2), are involved in the regulation of critical periods of brain development and neurogenesis (Planques et al., 2019; Prochiantz & Di Nardo, 2015). There is consensus that the amount and rates of adult neurogenesis in the mammalian brain and, specifically, the human brain are not ubiquitous and differ between brain areas and that this is related to different functional affordances (Emsley et al., 2005; Kempermann et al., 2018). For example, the human hippocampus has been observed to exhibit comparably high frequencies of adult neurogenesis (Bruel-Jungerman et al., 2007; Kempermann et al., 2018), with one possible function being the participation in the aforementioned representational drift in the context of memory updating (Mau et al., 2020).

The notion that new neurons evolve across the brain, or at least in dedicated brain areas, throughout the human lifespan not only sparks the question of why and how neural plasticity based on adult neurogenesis is facilitated but also why and how it is limited. The functional role of this limitation and the side effects for the single neuron need to be further considered, taking into account the associated energetic costs of psychobiological development and different mental functions. For example, in contrast to Mau et al. (2020) and their interpretation of adult neurogenesis as a mechanism of memory updating, Prochiantz (2012) holds the view that a permanently self-renewing CNS contradicts stable memory formation and that this is one important reason for its limitation in humans. However, both perspectives assume a strong functional stability of the single neuron, which, as discussed, does not hold in all cases. Again, the question of plasticity is strongly entangled in the tension between dynamic change and stabilization at the single-neuron level.

Before I discuss this further (see Chap. 3), I will introduce a last potential source of neural plasticity that specifically caters to the tension between stability and change. Taking the hint of homeoproteins, which are active within the nucleus and regulate gene expression and chromatin formation, we need to consider their impact on and interaction with epigenetic mechanisms within a single neuron and other brain cells. Epigenetic mechanisms are involved in adult neurogenesis in the mammalian brain (Ma et al., 2010). This is consistent with the view that epigenetic mechanisms play an important role in cell differentiation during development. However, this raises the question of whether epigenetic modifications add another layer to the neuron's plasticity.

Neuroepigenetic Mechanisms: Another Layer of Plasticity

Epigenetic mechanisms have been suggested to underlie a wide range of phenomena related to the plasticity of neuronal networks and cells. Edelstein and Smythies describe the "epigenetic code" as "dynamic hardware of the brain" (Edelstein & Smythies, 2014). Borelli et al. argue that epigenetic mechanisms, and especially those involved in chromatin remodeling, contribute tremendously to the "highly flexible nature of neurons" (Borrelli et al., 2008, p. 961). According to them, the interplay between genetic and epigenetic mechanisms underlies the intricate relationship between a stable cell memory and the ongoing integration of environmental cues, and this capability to integrate environmental signals is at the core of a neuron's functional tasks.

First evidence that epigenetic mechanisms contribute to function differently in neurons compared to other cells came from findings that DNA methyltransferases (DNMTs), specifically, DNMT1, DNMT3a, and DNMT3b, are expressed to a much higher degree in differentiated neurons than in other differentiated cells (Feng et al., 2005, 2010). Combined with knowledge about their intricate involvement in the regulation of gene expression, a potential functional role of epigenetic mechanisms in synaptic plasticity came into focus (Bayraktar et al., 2020; Bayraktar & Kreutz, 2018). The link between changes in gene expression patterns and synaptic plasticity is already well established. The increased expression of immediate early genes (IEGs), such as *c-fos* and *Arc*, after neural stimulation is a frequently used biomarker of synaptic activity and plasticity (Chowdhury et al., 2006; Chung, 2015; Durchdewald et al., 2009; Minatohara et al., 2016). In this context, those epigenetic mechanisms were of specific interest which modulate the chromatin structure in a way that affects—enhances or suppresses—transcription factor expression or binding of genes related to synaptic plasticity. This includes histone modifications and DNA methylation mainly and also extends to enzymatic processes involved in larger spatial chromatin remodeling (Campbell & Wood, 2019; Cortés-Mendoza et al., 2013; Gräff & Mansuy, 2008; Sultan & Day, 2011). Another indicator of a functional role of epigenetic mechanisms in synaptic plasticity is the participation of the CREB-binding protein (CBP/KAT3A), a histone acetyltransferase, in the CREB pathway involved in memory formation (Gräff & Tsai, 2013; Guan et al., 2002). Furthermore, hippocampal DNMT expression was found to increase after contextual fear conditioning, and DNMT inhibition blocked the formation of contextual fear memory (Miller & Sweatt, 2007). In line with these findings, epigenetic mechanisms that regulate synaptic function are considered to provide some of the molecular mechanisms underlying memory formation and mental health (Campbell & Wood, 2019; Day & Sweatt, 2010; Mews et al., 2021; Sultan & Day, 2011).

That epigenetic mechanisms specifically participate in the regulation of gene expression related to synaptic plasticity is mainly supported by studies investigating the role of epigenetic-modifying enzymes involved in histone modification and DNA de/methylation (Sultan & Day, 2011) and thus on a more general level. Observed changes are mostly short term, which is consistent with altered expression patterns following neural activation. This short-term flexibility and the observed dynamic patterns of histone modifications and DNA methylation were initially surprising (Day & Sweatt, 2010; Sultan & Day, 2011; Sweatt, 2013). They go against the original perception of epigenetic marks, and especially DNA methylation, as being a very stable mechanism of gene expression regulation, mainly involved in cell differentiation and the establishment of tissue-specific gene expression patterns, X chromosome inactivation, and imprinting (Sweatt, 2013). In addition, it is still heavily debated whether the transient epigenetic modifications observed in the context of memory formation and changes in synaptic plasticity markers have a specific functional role on their own or are merely a by-product of transcriptional feedback loops (Horsthemke, 2022). Nevertheless, histone modifications and DNA methylation in neurons have been repeatedly shown to be affected by neural activity (for review, see Campbell & Wood, 2019; Mews et al., 2021), and the inhibition of enzymes regulating them has been shown to affect memory formation (Chatterjee et al., 2013; Guan et al., 2002; Levenson et al., 2004; Miller et al., 2008). Both findings indicate that synaptic changes interact with the epigenetic machinery in a potentially functional manner. However, the picture is still unclear, and the search for the underlying molecular pathways is still ongoing.

Some findings point to an unspecific effect of synaptic activity on chromatin-modifying enzymes, in particular on the presence of histone acetyltransferases/deacetylases (HATs/HDACs), which interfere with the accessibility of transcription factors to the DNA by changing the chromatin density, and DNMTs involved in de novo DNA methylation of promoter regions and the recruitment of HDACs. For example, in vitro studies in hippocampal cell cultures indicate that acetylation and phosphorylation of histone H3 increase following activation of N-methyl-D-aspartate (NMDA) receptors and extracellular signal-regulated kinase (ERK) (Levenson et al., 2004). In addition, in vivo studies in rats showed a similar increase in histone H3 phosphorylation, likely regulated via the extracellular signal-regulated kinase/mitogen-activated protein kinase (ERK/MAPK) pathway (Chwang et al., 2006). Furthermore, Gadd45 (growth arrest and DNA-damage-inducible protein 45 alpha), a nuclear protein that is involved in the maintenance of genomic stability and DNA repair and has been found to participate in active DNA demethylation (Barreto et al., 2007), influences synaptic remodeling (Ma et al., 2010). Here, one hypothesis interprets these broad epigenome-wide effects as facilitating or priming mechanisms that activate or deactivate chromatin so that transcription loops of functionally relevant genes can be more easily adapted to the current transcriptional affordances.

In addition, however, to a smaller degree, some epigenetic changes have been observed to occur at genes related to synaptic plasticity, such as DNA methylation in the promoter regions of *reelin* and *bdnf* (Levenson et al., 2006) and histone H3 acetylation in the promoter region of *bdnf* (Chatterjee et al., 2013). In particular, the diverse epigenetic regulation of the brain-derived neurotropic factor (BDNF) seems to play a key role in the translation of transient changes into long-term changes with functional impact on neural plasticity (for a review see, for example, Kyrke-Smith & Williams, 2018). Furthermore, a possible role of genome-wide epigenetic modifications in non-Hebbian and homeostatic plasticity, independent of specific signaling pathways associated with synaptic plasticity, has been proposed (Kyrke-Smith & Williams, 2018; Sweatt, 2013). According to Sweatt (2013), epigenetic modifications facilitate some type of meta-plasticity by influencing cell-wide set-points for synapse-specific plasticity mechanisms and regulating a neuron's sensitivity, excitability, and activity-dependent synaptic scaling at a broad level. In glutamatergic neurons, a key mechanism for this function is the interaction between DNA methylation and glutamate receptor trafficking (Sweatt, 2016).

However, these mechanisms take part in a very small time window, starting with the synaptic input and lasting for up to 24 h after the first input. They participate in short-term transcriptional changes that underlie synaptic and homeostatic activity. Most of them are transient (as observed by Day & Sweatt, 2010 in the hippocampus), representing epigenetic mechanisms of short-term synaptic adaptation. Although they contribute to synaptic function, the question arises as to how these or other epigenetic mechanisms contribute to long-term changes in synaptic and neural network plasticity (see Campbell & Wood, 2019). Are these synaptic changes stabilized and maintained at the epigenome level? Or do epigenetic modifications only occur in relation to synaptic activity in a very short time window, and afterward, self-sustaining feedback loops at the synapse level take over? Here, it is important to distinguish between structural-genomic, synaptic, and developmental epigenetic mechanisms, as introduced in the last chapter. It allows for an analytical differentiation between epigenetic mechanisms related to ongoing homeostasis maintenance at the genome level, those transient changes related to synaptic activity, and long-term changes with relative stability that contribute to changes in developmental pathways at higher functional levels.

The same is the case for epigenetic modifications impacting synaptic plasticity that are not induced by neural activity but, for example, by hormones, neurotransmitters via neuromodulation, or direct cell-cell communication via exosomes. The impact of a variety of hormones, most importantly stress and sex hormones, on epigenetic mechanisms has been studied. A variety of epigenetic mechanisms are sensitive to the exposure to, for example, glucocorticoids (for review, see Lux, 2018) and testosterone (Bramble et al., 2016; Ghahramani et al., 2014). Hormones affect the epigenome mostly in an unspecific manner, but more specific functional adaptations resulting in behavioral differences have also been reported (for review, see Baumbach & Zovkic, 2020). One hypothesis is that similar to the unspecific epigenome-wide effects of synaptic activity, hormones, too, function as an unspecific primer for epigenetic modifications which either facilitates or stabilizes the functional specific epigenetic modifications initiated by one of the signaling pathways following neural activity ("dual-activation hypothesis," see Lux, 2018). In addition, the effect of hormones on epigenetic modifications has been reported to be stronger during critical developmental periods, for example, early in life, during periods of cell differentiation, or when hormonal regulation itself undergoes important changes (Bramble et al., 2016; Ghahramani et al., 2014). Thus, during these periods, the epigenome is more plastic in the sense that it is more sensitive to stimuli that facilitate adaptive changes.

Plasticity at the level of the epigenome thus means an increased likelihood of epigenetic modifications across the entire epigenome that are induced via the impact of neural activity on epigenetic functional enzymes or via hormones and that are enhanced during critical and sensitive developmental periods. In line with this notion are findings from cancer research (Flavahan, 2020; Flavahan et al., 2017). Here, epigenetic plasticity is understood as instability of the cell-specific homeostasis of the chromatin state in cancer cells, and this is discussed as one possible mechanism underlying cancer development. Such a loss of chromatin homeostasis results in a high variability of epigenetic modifications and stochastic gene expression patterns. This notion of epigenetic plasticity, however, contradicts the original view that epigenetic mechanisms provide an energy-efficient stabilization of cell-specific modifications in neurons and other brain cells. For a functionally specific connection between neural plasticity and epigenetic modifications, it seems to be more important how specific changes are stabilized at the epigenome level. Thus, in the case of memory formation, related epigenetic modifications need to function as a canalizing mechanism to meet the dedicated functional role as an energy-efficient stabilizing mechanism in memory formation. This includes specific mechanisms that need to be targeted, in contrast to the stochastic changes across the whole genome associated with cancer. Moreover, the maintenance of induced modifications beyond the immediate time point of neural activation is crucial. From an energetic point of view, stabilizing and maintaining induced changes at the single-neuron level are clearly the more complicated part.

Stabilizing Neural Plasticity: Three Characteristics of Developmental Epigenetic Mechanisms

In line with the earlier proposed functional differentiation of epigenetic mechanisms, I understand the mechanisms that stabilize long-term neuronal plasticity at the epigenetic level as developmental epigenetic mechanisms. This implicates three characteristics of these epigenetic mechanisms: First, in regard to the time scale, the definition of developmental epigenetic mechanisms implies that changes are medium- or long-term at the epigenetic level, in contrast to transient short-term changes related to general synaptic function and activity. Second, the respective developmental epigenetic modifications need to mark an enduring functional difference in the neuron's homeostatic equilibrium or activity patterns, which can be detected as distinct from the previous state. Third, these functional differences need to be related to functional or morphological changes involved in one or more higher levels of plasticity, such as synaptic, neuronal, cognitive, or behavioral plasticity, including memory and learning. According to these criteria, transient and stable epigenetic modifications related to fear memory would qualify as developmental epigenetic modifications that underlie plasticity. In contrast, global changes in DNA methylation due to increased glucocorticoid exposure or aging, which have not been linked to functional implications for a neuron's plasticity performance at one of the higher plasticity levels, would not qualify as such.

Although such analytic differentiation is crucial to differentiate epigenetic mechanisms with long-term impact on neural function and higher functional levels from more short-term and ongoing regulatory mechanisms, we always need to keep in mind that these processes occur together in a single neuron. Moreover, we must assume ongoing, direct, indirect, and partially stochastic interactions between structural-genomic, synaptic, and developmental epigenetic mechanisms at all times. Some interactions in the context of neural plasticity have been discussed to interpret available findings. For example, some synaptic epigenetic modifications that are associated with the regulation of immediate early gene expression in the context of neural activity are also involved in priming neurons and their network connections to stabilize a memory trace (Minatohara et al., 2016). In addition, structural-genomic epigenetic mechanisms implicated, for example, in the regulation of homeostatic plasticity, interact with transcriptional changes facilitating synaptic plasticity in learning and memory formation, and aging-related decline of learning and memory is related to a decline in homeostatic plasticity regulation (Kyrke-Smith & Williams, 2018).

However, from the perspective of the single neuron, developmental epigenetic mechanisms involved in the regulation of plasticity clearly stand out as making a difference, for example, in the range of functionality and functional integration at the neural network level, and in the neuron's involvement in behavior and higher psychological functions. This raises the question of how we capture these long-term neuroepigenetic changes and their relationship with plasticity at different functional levels.

References

Bailey, C. H., & Kandel, E. R. (1995). Molecular and structural mechanisms underlying long-term memory. In M. S. Gazzaniga (Ed.), *The cognitive neurosciences* (pp. 19–36). MIT Press.

Bailey, C. H., Bartsch, D., & Kandel, E. R. (1996). Toward a molecular definition of long-term memory storage. *Proceedings of the National Academy of Sciences, 93*(24), 13445–13452. https://doi.org/10.1073/pnas.93.24.13445

Barker, D. J. P. (2007). The origins of the developmental origins theory. *Journal of Internal Medicine, 261*(5), 412–417. https://doi.org/10.1111/j.1365-2796.2007.01809.x

Barreto, G., Schäfer, A., Marhold, J., Stach, D., Swaminathan, S. K., Handa, V., Döderlein, G., Maltry, N., Wu, W., Lyko, F., & Niehrs, C. (2007). Gadd45a promotes epigenetic gene activation by repair-mediated DNA demethylation. *Nature, 445*(7128), 671–675. https://doi.org/10.1038/nature05515

Bateson, P., & Gluckman, P. (2012). Plasticity and robustness in development and evolution. *International Journal of Epidemiology, 41*(1), 219–223. https://doi.org/10.1093/ije/dyr240

Baumbach, J. L., & Zovkic, I. B. (2020). Hormone-epigenome interactions in behavioural regulation. *Hormones and Behavior, 118*, 104680. https://doi.org/10.1016/j.yhbeh.2020.104680

Bayraktar, G., & Kreutz, M. R. (2018). Neuronal DNA methyltransferases: Epigenetic mediators between synaptic activity and gene expression? *The Neuroscientist, 24*(2), 171–185. https://doi.org/10.1177/1073858417707457

Bayraktar, G., Yuanxiang, P., Confettura, A. D., Gomes, G. M., Raza, S. A., Stork, O., Tajima, S., Suetake, I., Karpova, A., Yildirim, F., & Kreutz, M. R. (2020). Synaptic control of DNA methylation involves activity-dependent degradation of DNMT3A1 in the nucleus. *Neuropsychopharmacology, 45*(12), Article 12. https://doi.org/10.1038/s41386-020-0780-2

Bengoetxea, H., Ortuzar, N., Bulnes, S., Rico-Barrio, I., Lafuente, J. V., & Argandoña, E. G. (2012). Enriched and deprived sensory experience induces structural changes and rewires connectivity during the postnatal development of the brain. *Neural Plasticity, 2012*, e305693. https://doi.org/10.1155/2012/305693

Berlucchi, G., & Buchtel, H. A. (2009). Neuronal plasticity: Historical roots and evolution of meaning. *Experimental Brain Research, 192*(3), 307–319. https://doi.org/10.1007/s00221-008-1611-6

Bi, G., & Poo, M. (2001). Synaptic modification by correlated activity: Hebb's postulate revisited. *Annual Review of Neuroscience, 24*, 139–166. https://doi.org/10.1146/annurev.neuro.24.1.139

Bianco-Miotto, T., Craig, J. M., Gasser, Y. P., van Dijk, S. J., & Ozanne, S. E. (2017). Epigenetics and DOHaD: From basics to birth and beyond. *Journal of Developmental Origins of Health and Disease, 8*(5), 513–519. https://doi.org/10.1017/S2040174417000733

Bladon, J. H., Sheehan, D. J., Freitas, C. S. D., & Howard, M. W. (2019). In a temporally segmented experience hippocampal neurons represent temporally drifting context but not discrete segments. *Journal of Neuroscience, 39*(35), 6936–6952. https://doi.org/10.1523/JNEUROSCI.1420-18.2019

Borrelli, E., Nestler, E. J., Allis, C. D., & Sassone-Corsi, P. (2008). Decoding the epigenetic language of neuronal plasticity. *Neuron, 60*(6), 961–974. https://doi.org/10.1016/j.neuron.2008.10.012

Bouton, M. E., Maren, S., & McNally, G. P. (2021). Behavioral and neurobiological mechanisms of pavlovian and instrumental extinction learning. *Physiological Reviews, 101*(2), 611–681. https://doi.org/10.1152/physrev.00016.2020

Bramble, M. S., Roach, L., Lipson, A., Vashist, N., Eskin, A., Ngun, T., Gosschalk, J. E., Klein, S., Barseghyan, H., Arboleda, V. A., & Vilain, E. (2016). Sex-specific effects of testosterone on the sexually dimorphic transcriptome and epigenome of embryonic neural stem/progenitor cells. *Scientific Reports, 6*(1), Article 1. https://doi.org/10.1038/srep36916

Brod, G., Bunge, S. A., & Shing, Y. L. (2017). Does one year of schooling improve children's cognitive control and alter associated brain activation? *Psychological Science, 28*(7), 967–978. https://doi.org/10.1177/0956797617699838

Bruel-Jungerman, E., Rampon, C., & Laroche, S. (2007). Adult hippocampal neurogenesis, synaptic plasticity and memory: Facts and hypotheses. *Reviews in the Neurosciences, 18*(2), 93–114. https://doi.org/10.1515/revneuro.2007.18.2.93

Budd, A. M., Robins, J. B., Whybird, O., & Jerry, D. R. (2022). Epigenetics underpins phenotypic plasticity of protandrous sex change in fish. *Ecology and Evolution, 12*(3), e8730. https://doi.org/10.1002/ece3.8730

Bunge, S. A., & Leib, E. R. (2020). How does education hone reasoning ability? *Current Directions in Psychological Science, 29*(2), 167–173. https://doi.org/10.1177/0963721419898818

Campbell, R. R., & Wood, M. A. (2019). How the epigenome integrates information and reshapes the synapse. *Nature Reviews Neuroscience, 20*(3), 133–147. https://doi.org/10.1038/s41583-019-0121-9

Campelo, T., Augusto, E., Chenouard, N., de Miranda, A., Kouskoff, V., Camus, C., Choquet, D., & Gambino, F. (2020). AMPAR-dependent synaptic plasticity initiates cortical remapping and adaptive behaviors during sensory experience. *Cell Reports, 32*(9), 108097. https://doi.org/10.1016/j.celrep.2020.108097

Chambers, A. R., & Rumpel, S. (2017). A stable brain from unstable components: Emerging concepts and implications for neural computation. *Neuroscience, 357*, 172–184. https://doi.org/10.1016/j.neuroscience.2017.06.005

Chatterjee, S., Mizar, P., Cassel, R., Neidl, R., Selvi, B. R., Mohankrishna, D. V., Vedamurthy, B. M., Schneider, A., Bousiges, O., Mathis, C., Cassel, J.-C., Eswaramoorthy, M., Kundu, T. K., & Boutillier, A.-L. (2013). A novel activator of CBP/p300 acetyltransferases promotes neurogenesis and extends memory duration in adult mice. *Journal of Neuroscience, 33*(26), 10698–10712. https://doi.org/10.1523/JNEUROSCI.5772-12.2013

Choe, Y. (2014). Anti-Hebbian learning. In D. Jaeger & R. Jung (Eds.), *Encyclopedia of computational neuroscience* (pp. 1–4). Springer. https://doi.org/10.1007/978-1-4614-7320-6_675-1

Chowdhury, S., Shepherd, J. D., Okuno, H., Lyford, G., Petralia, R. S., Plath, N., Kuhl, D., Huganir, R. L., & Worley, P. F. (2006). Arc/Arg3.1 interacts with the endocytic machinery to regulate AMPA receptor trafficking. *Neuron, 52*(3), 445–459. https://doi.org/10.1016/j.neuron.2006.08.033

Chung, L. (2015). A brief introduction to the transduction of neural activity into Fos signal. *Development & Reproduction, 19*(2), 61–67. https://doi.org/10.12717/DR.2015.19.2.061

Chwang, W. B., O'Riordan, K. J., Levenson, J. M., & Sweatt, J. D. (2006). ERK/MAPK regulates hippocampal histone phosphorylation following contextual fear conditioning. *Learning & Memory, 13*(3), 322–328. https://doi.org/10.1101/lm.152906

Citri, A., & Malenka, R. C. (2008). Synaptic plasticity: Multiple forms, functions, and mechanisms. *Neuropsychopharmacology, 33*(1), Article 1. https://doi.org/10.1038/sj.npp.1301559

Clopath, C., Bonhoeffer, T., Hübener, M., & Rose, T. (2017). Variance and invariance of neuronal long-term representations. *Philosophical Transactions of the Royal Society B: Biological Sciences, 372*(1715), 20160161. https://doi.org/10.1098/rstb.2016.0161

Commins, S. (2018). Habituation and sensitisation in the Aplysia. In *Behavioural neuroscience* (pp. 48–61). Cambridge University Press. https://doi.org/10.1017/9781316221655.006

Corner, M. A., & Ramakers, G. J. (1992). Spontaneous firing as an epigenetic factor in brain development—Physiological consequences of chronic tetrodotoxin and picrotoxin exposure on cultured rat neocortex neurons. *Developmental Brain Research, 65*(1), 57–64. https://doi.org/10.1016/0165-3806(92)90008-k

Cortés-Mendoza, J., Díaz de León-Guerrero, S., Pedraza-Alva, G., & Pérez-Martínez, L. (2013). Shaping synaptic plasticity: The role of activity-mediated epigenetic regulation on gene transcription. *International Journal of Developmental Neuroscience: The Official Journal of the International Society for Developmental Neuroscience, 31*(6), 359–369. https://doi.org/10.1016/j.ijdevneu.2013.04.003

Das, A. (1997). Plasticity in adult sensory cortex: A review. *Network: Computation in Neural Systems, 8*(2), R33–R76. https://doi.org/10.1088/0954-898X_8_2_001

Day, J. J., & Sweatt, J. D. (2010). DNA methylation and memory formation. *Nature Neuroscience, 13*(11), 1319–1323. https://doi.org/10.1038/nn.2666

Dunsmoor, J. E., Niv, Y., Daw, N., & Phelps, E. A. (2015). Rethinking extinction. *Neuron, 88*(1), 47–63. https://doi.org/10.1016/j.neuron.2015.09.028

Durchdewald, M., Angel, P., & Hess, J. (2009). The transcription factor Fos: A Janus-type regulator in health and disease. *Histology and Histopathology, 24*(11), 1451–1461. https://doi.org/10.14670/HH-24.1451

Edelstein, L., & Smythies, J. (2014). The role of epigenetic-related codes in neurocomputation: Dynamic hardware in the brain. *Philosophical Transactions of the Royal Society B: Biological Sciences, 369*(1652), 20130519. https://doi.org/10.1098/rstb.2013.0519

Elbert, T., Pantev, C., Wienbruch, C., Rockstroh, B., & Taub, E. (1995). Increased cortical representation of the fingers of the left hand in string players. *Science (New York, N.Y.), 270*(5234), 305–307. https://doi.org/10.1126/science.270.5234.305

Emsley, J. G., Mitchell, B. D., Kempermann, G., & Macklis, J. D. (2005). Adult neurogenesis and repair of the adult CNS with neural progenitors, precursors, and stem cells. *Progress in Neurobiology, 75*(5), 321–341. https://doi.org/10.1016/j.pneurobio.2005.04.002

Erzurumlu, R. S., & Gaspar, P. (2020). How the barrel cortex became a working model for developmental plasticity: A historical perspective. *Journal of Neuroscience, 40*(34), 6460–6473. https://doi.org/10.1523/JNEUROSCI.0582-20.2020

Espinosa, J. S., & Stryker, M. P. (2012). Development and plasticity of the primary visual cortex. *Neuron, 75*(2), 230–249. https://doi.org/10.1016/j.neuron.2012.06.009

Feng, J., Chang, H., Li, E., & Fan, G. (2005). Dynamic expression of de novo DNA methyltransferases Dnmt3a and Dnmt3b in the central nervous system. *Journal of Neuroscience Research, 79*(6), 734–746. https://doi.org/10.1002/jnr.20404

Feng, J., Zhou, Y., Campbell, S. L., Le, T., Li, E., Sweatt, J. D., Silva, A. J., & Fan, G. (2010). Dnmt1 and Dnmt3a maintain DNA methylation and regulate synaptic function in adult forebrain neurons. *Nature Neuroscience, 13*(4), Article 4. https://doi.org/10.1038/nn.2514

Flament, S. (2016). Sex reversal in amphibians. *Sexual Development, 10*(5–6), 267–278. https://doi.org/10.1159/000448797

Flavahan, W. A. (2020). Epigenetic plasticity, selection, and tumorigenesis. *Biochemical Society Transactions, 48*(4), 1609–1621. https://doi.org/10.1042/BST20191215

Flavahan, W. A., Gaskell, E., & Bernstein, B. E. (2017). Epigenetic plasticity and the hallmarks of cancer. *Science, 357*(6348), eaal2380. https://doi.org/10.1126/science.aal2380

Franks, K. M., & Isaacson, J. S. (2005). Synapse-specific downregulation of NMDA receptors by early experience: A critical period for plasticity of sensory input to olfactory cortex. *Neuron, 47*(1), 101–114. https://doi.org/10.1016/j.neuron.2005.05.024

Ghahramani, N. M., Ngun, T. C., Chen, P.-Y., Tian, Y., Krishnan, S., Muir, S., Rubbi, L., Arnold, A. P., de Vries, G. J., Forger, N. G., Pellegrini, M., & Vilain, E. (2014). The effects of perinatal testosterone exposure on the DNA methylome of the mouse brain are late-emerging. *Biology of Sex Differences, 5*(1), 8. https://doi.org/10.1186/2042-6410-5-8

Gilbert, S. F. (2001). Ecological developmental biology: Developmental biology meets the real world. *Developmental Biology, 233*(1), 1–12. https://doi.org/10.1006/dbio.2001.0210

Gilbert, S. F., & Barresi, M. J. F. (2020). *Developmental biology* (12th ed.). Oxford University Press.

Gilbert, S. F., Bosch, T. C. G., & Ledón-Rettig, C. (2015). Eco-Evo-Devo: Developmental symbiosis and developmental plasticity as evolutionary agents. *Nature Reviews Genetics, 16*(10), 611–622. https://doi.org/10.1038/nrg3982

Gluckman, P. D., Hanson, M. A., Spencer, H. G., & Bateson, P. (2005). Environmental influences during development and their later consequences for health and disease: Implications for the interpretation of empirical studies. *Proceedings. Biological Sciences, 272*(1564), 671–677. https://doi.org/10.1098/rspb.2004.3001

Goyal, D., Limesand, S. W., & Goyal, R. (2019). Epigenetic responses and the developmental origins of health and disease. *Journal of Endocrinology, 242*(1), T105–T119. https://doi.org/10.1530/JOE-19-0009

Gräff, J., & Mansuy, I. M. (2008). Epigenetic codes in cognition and behaviour. *Behavioural Brain Research, 192*(1), 70–87. https://doi.org/10.1016/j.bbr.2008.01.021

Gräff, J., & Tsai, L.-H. (2013). Histone acetylation: Molecular mnemonics on the chromatin. *Nature Reviews. Neuroscience, 14*(2), 97–111. https://doi.org/10.1038/nrn3427

Greenough, W., Black, J., & Wallace, C. (1987). Experience and brain development. *Child Development, 58*(3), 539–559. https://pubmed.ncbi.nlm.nih.gov/3038480/

Guan, Z., Giustetto, M., Lomvardas, S., Kim, J.-H., Miniaci, M. C., Schwartz, J. H., Thanos, D., & Kandel, E. R. (2002). Integration of long-term-memory-related synaptic plasticity involves bidirectional regulation of gene expression and chromatin structure. *Cell, 111*(4), 483–493. https://doi.org/10.1016/s0092-8674(02)01074-7

Guerra-Carrillo, B., Mackey, A. P., & Bunge, S. A. (2014). Resting-state fMRI: A window into human brain plasticity. *The Neuroscientist: A Review Journal Bringing Neurobiology, Neurology and Psychiatry, 20*(5), 522–533. https://doi.org/10.1177/1073858414524442

Guic, E., Carrasco, X., Rodríguez, E., Robles, I., & Merzenich, M. M. (2008). Plasticity in primary somatosensory cortex resulting from environmentally enriched stimulation and sensory discrimination training. *Biological Research, 41*(4), 425–437. https://doi.org/10.4067/S0716-97602008000400008

Hagmann, P. (2005). *From diffusion MRI to brain connectomics* (Thèse no 3230). EPFL, Lausanne. https://doi.org/10.5075/epfl-thesis-3230

Hagmann, P., Cammoun, L., Gigandet, X., Gerhard, S., Grant, P. E., Wedeen, V., Meuli, R., Thiran, J.-P., Honey, C. J., & Sporns, O. (2010). MR connectomics: Principles and challenges. *Journal of Neuroscience Methods, 194*(1), 34–45. https://doi.org/10.1016/j.jneumeth.2010.01.014

Hales, C. N., & Barker, D. J. (1992). Type 2 (non-insulin-dependent) diabetes mellitus: The thrifty phenotype hypothesis. *Diabetologia, 35*(7), 595–601. https://doi.org/10.1007/BF00400248

Hansel, C., Linden, D. J., & D'Angelo, E. (2001). Beyond parallel fiber LTD: The diversity of synaptic and non-synaptic plasticity in the cerebellum. *Nature Neuroscience, 4*(5), 467–475. https://doi.org/10.1038/87419

Hanson, M. A., & Gluckman, P. D. (2014). Early developmental conditioning of later health and disease: Physiology or pathophysiology? *Physiological Reviews, 94*(4), 1027–1076. https://doi.org/10.1152/physrev.00029.2013

Hebb, D. O. (1949). *The organization of behavior: A neuropsychological theory.* John Wiley & Sons.

Hooks, B. M., & Chen, C. (2007). Critical periods in the visual system: Changing views for a model of experience-dependent plasticity. *Neuron, 56*(2), 312–326. https://doi.org/10.1016/j.neuron.2007.10.003

Horsthemke, B. (2022). A critical appraisal of clinical epigenetics. *Clinical Epigenetics, 14*(1), 95. https://doi.org/10.1186/s13148-022-01315-6

Hsu, C.-N., & Tain, Y.-L. (2021). Animal models for DOHaD research: Focus on hypertension of developmental origins. *Biomedicine, 9*(6), 623. https://doi.org/10.3390/biomedicines9060623

Hubel, D. H., & Wiesel, T. N. (1963). Receptive fields of cells in striate cortex of very young, visually inexperienced kittens. *Journal of Neurophysiology, 26*, 994–1002.

James, W. (1890). *The principles of psychology* (Vol. 1). Henry Holt and Company.

Johnson, M. H. (2011). Interactive specialization: A domain-general framework for human functional brain development? *Developmental Cognitive Neuroscience, 1*(1), 7–21. https://doi.org/10.1016/j.dcn.2010.07.003

Johnson, C., Kretsge, L. N., Yen, W. W., Sriram, B., O'Connor, A., Liu, R. S., et al. (2022). Highly unstable heterogeneous representations in VIP interneurons of the anterior cingulate cortex. *Molecular Psychiatry, 27*(5), 2602–2618. https://doi.org/10.1038/s41380-022-01485-y

Kaiser, M., & Cromby, J. (2014). Neuroscience. In T. Teo (Ed.), *Encyclopedia of critical psychology* (pp. 1243–1248). Springer. https://doi.org/10.1007/978-1-4614-5583-7_200

Kandel, E. R. (2000). *The molecular biology of memory storage: A dialogue between genes and synapses. Nobel lecture.* https://www.nobelprize.org/uploads/2018/06/kandel-lecture.pdf

Keck, T., Toyoizumi, T., Chen, L., Doiron, B., Feldman, D. E., Fox, K., Gerstner, W., Haydon, P. G., Hübener, M., Lee, H.-K., Lisman, J. E., Rose, T., Sengpiel, F., Stellwagen, D., Stryker, M. P., Turrigiano, G. G., & van Rossum, M. C. (2017). Integrating Hebbian and homeostatic plasticity: The current state of the field and future research directions. *Philosophical Transactions of the Royal Society of London. Series B, Biological Sciences, 372*(1715), 20160158. https://doi.org/10.1098/rstb.2016.0158

Keller, L., & Ross, K. G. (1993). Phenotypic plasticity and "cultural transmission" of alternative social organizations in the fire ant Solenopsis invicta. *Behavioral Ecology and Sociobiology, 33*(2), 121–129. https://doi.org/10.1007/BF00171663

Kemenes, I., Straub, V. A., Nikitin, E. S., Staras, K., O'Shea, M., Kemenes, G., & Benjamin, P. R. (2006). Role of delayed nonsynaptic neuronal plasticity in long-term associative memory. *Current Biology: CB, 16*(13), 1269–1279. https://doi.org/10.1016/j.cub.2006.05.049

Kempermann, G., Gage, F. H., Aigner, L., Song, H., Curtis, M. A., Thuret, S., Kuhn, H. G., Jessberger, S., Frankland, P. W., Cameron, H. A., Gould, E., Hen, R., Abrous, D. N., Toni, N., Schinder, A. F., Zhao, X., Lucassen, P. J., & Frisén, J. (2018). Human adult neurogenesis: Evidence and remaining questions. *Cell Stem Cell, 23*(1), 25–30. https://doi.org/10.1016/j.stem.2018.04.004

Klingenberg, C. P. (2019). Phenotypic plasticity, developmental instability, and robustness: The concepts and how they are connected. *Frontiers in Ecology and Evolution, 7*, 56. https://www.frontiersin.org/articles/10.3389/fevo.2019.00056

Knudsen, E. I., Knudsen, P. F., & Esterly, S. D. (1984). A critical period for the recovery of sound localization accuracy following monaural occlusion in the barn owl. *The Journal of Neuroscience: The Official Journal of the Society for Neuroscience, 4*(4), 1012–1020.

Koelsch, S., Schröger, E., & Tervaniemi, M. (1999). Superior pre-attentive auditory processing in musicians. *Neuroreport, 10*(6), 1309–1313. https://doi.org/10.1097/00001756-199904260-00029

Krüger, M., & Lux, V. (2023). Failure of motor function—A Developmental Embodiment Research perspective on the systemic effects of stress. *Frontiers in Human Neuroscience, 17*, 1083200. https://www.frontiersin.org/articles/10.3389/fnhum.2023.1083200

Kullmann, D. M., & Lamsa, K. P. (2007). Long-term synaptic plasticity in hippocampal interneurons. *Nature Reviews. Neuroscience, 8*(9), 687–699. https://doi.org/10.1038/nrn2207

Kyrke-Smith, M., & Williams, J. M. (2018). Bridging synaptic and epigenetic maintenance mechanisms of the engram. *Frontiers in Molecular Neuroscience, 11*, 369. https://www.frontiersin.org/articles/10.3389/fnmol.2018.00369

Lazaris, C., Aifantis, I., & Tsirigos, A. (2020). On epigenetic plasticity and genome topology. *Trends in Cancer, 6*(3), 177–180. https://doi.org/10.1016/j.trecan.2020.01.006

Levenson, J. M., O'Riordan, K. J., Brown, K. D., Trinh, M. A., Molfese, D. L., & Sweatt, J. D. (2004). Regulation of histone acetylation during memory formation in the hippocampus. *Journal of Biological Chemistry, 279*(39), 40545–40559. https://doi.org/10.1074/jbc.M402229200

Levenson, J. M., Roth, T. L., Lubin, F. D., Miller, C. A., Huang, I.-C., Desai, P., Malone, L. M., & Sweatt, J. D. (2006). Evidence that DNA (cytosine-5) methyltransferase regulates synaptic plasticity in the hippocampus. *The Journal of Biological Chemistry, 281*(23), 15763–15773. https://doi.org/10.1074/jbc.M511767200

Liberti, W. A., Markowitz, J. E., Perkins, L. N., Liberti, D. C., Leman, D. P., Guitchounts, G., Velho, T., Kotton, D. N., Lois, C., & Gardner, T. J. (2016). Unstable neurons underlie a stable learned behavior. *Nature Neuroscience, 19*(12), 1665–1671. https://doi.org/10.1038/nn.4405

Lux, V. (2018). Epigenetic programming effects of early life stress: A dual-activation hypothesis. *Current Genomics, 19*(8), 638–652. https://doi.org/10.2174/1389202919666180307151358

Ma, D. K., Marchetto, M. C., Guo, J. U., Ming, G., Gage, F. H., & Song, H. (2010). Epigenetic choreographers of neurogenesis in the adult mammalian brain. *Nature Neuroscience, 13*(11), 1338–1344. https://doi.org/10.1038/nn.2672

Mackey, A. P., Singley, A. T. M., & Bunge, S. A. (2013). Intensive reasoning training alters patterns of brain connectivity at rest. *Journal of Neuroscience, 33*(11), 4796–4803. https://doi.org/10.1523/JNEUROSCI.4141-12.2013

Maguire, E. A., Gadian, D. G., Johnsrude, I. S., Good, C. D., Ashburner, J., Frackowiak, R. S. J., & Frith, C. D. (2000). Navigation-related structural change in the hippocampi of taxi drivers. *Proceedings of the National Academy of Sciences, 97*(8), 4398–4403. https://doi.org/10.1073/pnas.070039597

Manuck, S. B. (2010). The reaction norm in gene-environment interaction. *Molecular Psychiatry, 15*(9), 881–882. https://doi.org/10.1038/mp.2009.139

Margolis, D. J., Lütcke, H., Schulz, K., Haiss, F., Weber, B., Kügler, S., Hasan, M. T., & Helmchen, F. (2012). Reorganization of cortical population activity imaged throughout long-term sensory deprivation. *Nature Neuroscience, 15*(11), Article 11. https://doi.org/10.1038/nn.3240

Mau, W., Hasselmo, M. E., & Cai, D. J. (2020). The brain in motion: How ensemble fluidity drives memory-updating and flexibility. *eLife, 9*, e63550. https://doi.org/10.7554/eLife.63550

McMullen, S., & Mostyn, A. (2009). Animal models for the study of the developmental origins of health and disease: Workshop on "Nutritional models of the developmental origins of adult health and disease". *Proceedings of the Nutrition Society, 68*(3), 306–320. https://doi.org/10.1017/S0029665109001396

Merzenich, M. M. (2013). *Soft-wired: How the new science of brain plasticity can change your life* (2nd ed.). Parnassus Publ.

Merzenich, M. M., & DeCharms, R. C. (1996). Neural representations, experience, and change. In R. Llinas & P. Chruchland (Eds.), *The mind-brain continuum* (pp. 61–81). MIT Press.

Merzenich, M. M., & Jenkins, W. M. (1993). Reorganization of cortical representations of the hand following alterations of skin inputs induced by nerve injury, skin island transfers, and experience. *Journal of Hand Therapy: Official Journal of the American Society of Hand Therapists, 6*(2), 89–104. https://doi.org/10.1016/s0894-1130(12)80290-0

Mews, P., Calipari, E. S., Day, J., Lobo, M. K., Bredy, T., & Abel, T. (2021). From circuits to chromatin: The emerging role of epigenetics in mental health. *Journal of Neuroscience, 41*(5), 873–882. https://doi.org/10.1523/JNEUROSCI.1649-20.2020

Miller, C. A., & Sweatt, J. D. (2007). Covalent modification of DNA regulates memory formation. *Neuron, 53*(6), 857–869. https://doi.org/10.1016/j.neuron.2007.02.022

Miller, C. A., Campbell, S. L., & Sweatt, J. D. (2008). DNA methylation and histone acetylation work in concert to regulate memory formation and synaptic plasticity. *Neurobiology of Learning and Memory, 89*(4), 599–603. https://doi.org/10.1016/j.nlm.2007.07.016

Minatohara, K., Akiyoshi, M., & Okuno, H. (2016). Role of immediate-early genes in synaptic plasticity and neuronal ensembles underlying the memory trace. *Frontiers in Molecular Neuroscience, 8*, 78. https://www.frontiersin.org/articles/10.3389/fnmol.2015.00078

Mozzachiodi, R., & Byrne, J. H. (2010). More than synaptic plasticity: Role of nonsynaptic plasticity in learning and memory. *Trends in Neurosciences, 33*(1), 17. https://doi.org/10.1016/j.tins.2009.10.001

Münte, T. F., Altenmüller, E., & Jäncke, L. (2002). The musician's brain as a model of neuroplasticity. *Nature Reviews Neuroscience, 3*(6), Article 6. https://doi.org/10.1038/nrn843

Nahum, M., Lee, H., & Merzenich, M. M. (2013). Principles of neuroplasticity-based rehabilitation. *Progress in Brain Research, 207*, 141–171. https://doi.org/10.1016/B978-0-444-63327-9.00009-6

Nottebohm, F. (2005). The neural basis of birdsong. *PLoS Biology, 3*(5), e164. https://doi.org/10.1371/journal.pbio.0030164

Odling-Smee, F. J., Laland, K. N., & Feldman, M. W. (2003). *Niche construction: The neglected process in evolution* (Online-ausg). Princeton University Press.

Oyama, S. (2000). *The ontogeny of information: Developmental systems and evolution* (2nd ed., rev. and expanded). Duke University Press.

Pawela, C., & Biswal, B. (2011). Brain connectivity: A new journal emerges. *Brain Connectivity, 1*(1), 1–2. https://doi.org/10.1089/brain.2011.0020

Peters, A. J., Lee, J., Hedrick, N. G., O'Neil, K., & Komiyama, T. (2017). Reorganization of corticospinal output during motor learning. *Nature Neuroscience, 20*(8), Article 8. https://doi.org/10.1038/nn.4596

Piersma, T., & Drent, J. (2003). Phenotypic flexibility and the evolution of organismal design. *Trends in Ecology & Evolution, 18*(5), 228–233. https://doi.org/10.1016/S0169-5347(03)00036-3

Planques, A., Oliveira Moreira, V., Dubreuil, C., Prochiantz, A., & Di Nardo, A. A. (2019). OTX2 signals from the choroid plexus to regulate adult neurogenesis. *ENeuro, 6*(2), ENEURO.0262-18.2019. https://doi.org/10.1523/ENEURO.0262-18.2019

Prochiantz, A. (2012). *Qu'est-ce que le vivant?* Éditions du Seuil.

Prochiantz, A., & Di Nardo, A. A. (2015). Homeoprotein signaling in the developing and adult nervous system. *Neuron, 85*(5), 911–925. https://doi.org/10.1016/j.neuron.2015.01.019

Prull, M. W., Gabrieli, J. D. E., & Bunge, S. A. (2000). Age-related changes in memory: A cognitive neuroscience perspective. In *The handbook of aging and cognition* (2nd ed., pp. 91–153). Lawrence Erlbaum Associates Publishers.

Rees, T. (2016). *Plastic reason: An anthropology of brain science in embryogenetic terms.* University of California Press.

Ribic, A. (2020). Stability in the face of change: Lifelong experience-dependent plasticity in the sensory cortex. *Frontiers in Cellular Neuroscience, 14*, 76. https://www.frontiersin.org/articles/10.3389/fncel.2020.00076

Rüsseler, J., Altenmüller, E., Nager, W., Kohlmetz, C., & Münte, T. F. (2001). Event-related brain potentials to sound omissions differ in musicians and non-musicians. *Neuroscience Letters, 308*(1), 33–36. https://doi.org/10.1016/s0304-3940(01)01977-2

Scheler, G., & Fellous, J.-M. (2001). Dopamine modulation of prefrontal delay activity-reverberatory activity and sharpness of tuning curves. *Neurocomputing, 38–40*, 1549–1556. https://doi.org/10.1016/S0925-2312(01)00559-8

Seidel-Marzi, O., & Ragert, P. (2020). Neurodiagnostics in sports: Investigating the athlete's brain to augment performance and sport-specific skills. *Frontiers in Human Neuroscience, 14*, 133. https://doi.org/10.3389/fnhum.2020.00133

Sporns, O., Tononi, G., & Kötter, R. (2005). The human connectome: A structural description of the human brain. *PLoS Computational Biology, 1*(4), e42. https://doi.org/10.1371/journal.pcbi.0010042

Stotz, K. (2017). Why developmental niche construction is not selective niche construction: And why it matters. *Interface Focus, 7*(5), 20160157. https://doi.org/10.1098/rsfs.2016.0157

Sultan, F. A., & Day, J. J. (2011). Epigenetic mechanisms in memory and synaptic function. *Epigenomics, 3*(2), 157–181. https://doi.org/10.2217/epi.11.6

Sweatt, J. D. (2013). The emerging field of neuroepigenetics. *Neuron, 80*(3), 624–632. https://doi.org/10.1016/j.neuron.2013.10.023

Sweatt, J. D. (2016). Dynamic DNA methylation controls glutamate receptor trafficking and synaptic scaling. *Journal of Neurochemistry, 137*(3), 312–330. https://doi.org/10.1111/jnc.13564

Todd, E. V., Liu, H., Muncaster, S., & Gemmell, N. J. (2016). Bending genders: The biology of natural sex change in fish. *Sexual Development, 10*(5–6), 223–241. https://doi.org/10.1159/000449297

Tsao, A., Sugar, J., Lu, L., Wang, C., Knierim, J. J., Moser, M.-B., & Moser, E. I. (2018). Integrating time from experience in the lateral entorhinal cortex. *Nature, 561*(7721), Article 7721. https://doi.org/10.1038/s41586-018-0459-6

Turrigiano, G. G. (2017). The dialectic of Hebb and homeostasis. *Philosophical Transactions of the Royal Society of London. Series B, Biological Sciences, 372*(1715), 20160258. https://doi.org/10.1098/rstb.2016.0258

Turrigiano, G. G., & Nelson, S. B. (2000). Hebb and homeostasis in neuronal plasticity. *Current Opinion in Neurobiology, 10*(3), 358–364. https://doi.org/10.1016/s0959-4388(00)00091-x

Turrigiano, G. G., Leslie, K. R., Desai, N. S., Rutherford, L. C., & Nelson, S. B. (1998). Activity-dependent scaling of quantal amplitude in neocortical neurons. *Nature, 391*(6670), 892–896. https://doi.org/10.1038/36103

Wendelken, C., Ferrer, E., Whitaker, K. J., & Bunge, S. A. (2016). Fronto-parietal network reconfiguration supports the development of reasoning ability. *Cerebral Cortex (New York, N.Y.: 1991), 26*(5), 2178–2190. https://doi.org/10.1093/cercor/bhv050

West-Eberhard, M. J. (2003). *Developmental plasticity and evolution.* Oxford University Press.

West-Eberhard, M. J. (2005). Developmental plasticity and the origin of species differences. *Proceedings of the National Academy of Sciences, 102*(suppl_1), 6543–6549. https://doi.org/10.1073/pnas.0501844102

Woltereck, R. (1913). Weitere experimentelle untersuchungen über Artänderung, speziell über das Wesen quantitativer Artunterschiede bei Daphniden. *Zeitschrift für Induktive Abstammungs-und Vererbungslehre, 9*(1), 146–146. https://doi.org/10.1007/BF01876686

Yamada, L., & Chong, S. (2017). Epigenetic studies in developmental origins of health and disease: Pitfalls and key considerations for study design and interpretation. *Journal of Developmental Origins of Health and Disease, 8*(1), 30–43. https://doi.org/10.1017/S2040174416000507

Yin, W., Li, T., Hung, S.-C., Zhang, H., Wang, L., Shen, D., Zhu, H., Mucha, P. J., Cohen, J. R., & Lin, W. (2020). The emergence of a functionally flexible brain during early infancy. *Proceedings of the National Academy of Sciences, 117*(38), 23904–23913. https://doi.org/10.1073/pnas.2002645117

Zhang, W., & Linden, D. J. (2003). The other side of the engram: Experience-driven changes in neuronal intrinsic excitability. *Nature Reviews Neuroscience, 4*(11), Article 11. https://doi.org/10.1038/nrn1248

Zhao, T. C., Llanos, F., Chandrasekaran, B., & Kuhl, P. K. (2022). Language experience during the sensitive period narrows infants' sensory encoding of lexical tones-Music intervention reverses it. *Frontiers in Human Neuroscience, 16*, 941853. https://doi.org/10.3389/fnhum.2022.941853

Ziemann, U., Ilić, T. V., & Jung, P. (2006). Chapter 3: Long-term potentiation (LTP)-like plasticity and learning in human motor cortex – Investigations with transcranial magnetic stimulation (TMS). In C. Barber, S. Tsuji, S. Tobimatsu, T. Uozumi, N. Akamatsu, & A. Eisen (Eds.), *Supplements to clinical neurophysiology* (Vol. 59, pp. 19–25). Elsevier. https://doi.org/10.1016/S1567-424X(09)70007-8

Chapter 3
Canalizing Change

> This is pure speculation, but if we accept the idea that the process of individualization consists of adding memory on memory, that means keeping the register open but without erasing what has already been inscribed […], and if we admit that this infinite openness, which also builds on past experiences, is a mode of individual adaptation, then we could conclude that the price we pay for this form of adaptation are the breaks we push on the plasticity we encounter in other organisms. (Prochiantz, 2012, p. 151; translation: V.L.)[1]

In his philosophical book *Quest-ce que le vivant?* (What is living matter?, 2012), the neurobiologist Alain Prochiantz reflects on the observation that humans exhibit a substantially lower degree of adult neurogenesis than other species. He discusses that the immense human capability of individual adaptation and learning—which he interprets as process of "individualization"—requires an infinite process of memory building, and he then speculates that the reason for the brain's limited plasticity, and specifically the reduced rates of adult neurogenesis, may be that this tremendous memory capacity requires stabilization at the neural network and single-cell level (Prochiantz, 2012, p. 151). According to this notion, it would be the *constraint* of plasticity at the level of the single neuron, which enables a large range of individual variations and plasticity in human behavior.

In artificial neural networks, the notion that function emerges from reduced plasticity and cellular specialization is successfully used to design machine learning algorithms that model real-time behavioral changes and task learning. These algorithms enable artificial systems to act faster and more adequately in certain scenarios and tasks due to experience-based adjustments of the probabilities of a desired outcome. With every input step by step, they reduce the possible pathways to reach the goal. However, when we model learning as a process through which the multiplicity of outcomes, action pathways, or functional connections within a network is reduced stepwise until it reaches a final steady state, how can such a network

---

[1] "Nous sommes ici en plein spéculation, mais si nous acceptons l'idée que le processus d'individuation consiste à ajouter de la mémoire à de la mémoire, c'est-à-dire à laisser le register ouvert, mais sans effacer ce qui est déjà écrit […], et si nous admettons que cette ouverture vers l'indéterminé qui s'appuie aussi sur l'expérience passé est un modalité d'adaptation individuelle, alors on pourra en conclure que le prix à payer pour cette forme d'adaptation est bien le frein mis à la plasticité rencontrée chez d'autres organismes." (Prochiantz, 2012, p. 151)

© The Author(s), under exclusive license to Springer Nature Switzerland AG 2024
V. Lux, *The Neuron in Context*, SpringerBriefs in Psychology,
https://doi.org/10.1007/978-3-031-55229-8_3

continue learning when the perfect solution is found and the task is learned? Accordingly, exposing such a steady-state network to continued external disturbances can cause the entire system to experience "catastrophic forgetting" (Kirkpatrick et al., 2017; Parisi et al., 2019). In such an instance, the system collapses and all previous experiences and learned patterns are invalidated. This phenomenon is part of the stability–plasticity dilemma in self-learning computational networks (Mermillod et al., 2013). How can a system of defined endless connections continue learning after it reaches an optimal steady state without having to start all over again? Interestingly, the few computational solutions for the stability–plasticity dilemma in artificial neural networks focus on two dimensions: they either artificially introduce fuzziness in the learning process or in the set goals, including partial forgetting, which keeps the system from reaching a steady state, or they establish hierarchically structured levels of interdependence with varying degrees of freedom (Mermillod et al., 2013; Muñoz-Martin et al., 2021; Parisi et al., 2019; Schug et al., 2021).

In his reflections on the role of adult neurogenesis in the human brain, Prochiantz also proposed a combination of hierarchical structure and partial forgetting to explain the role of adult neurogenesis in memory formation. According to this notion, adult neurogenesis in the human hippocampus ensures that memories, which are already stored in the cortex and should not be kept longer at the level of the hippocampus, are erased (Prochiantz, 2012, p. 76). While functional specialization is maintained at the single-neuron level, this plasticity is ensured via the inclusion of new neurons that replace the already specialized neurons in the respective neural networks. This notion of plasticity goes against the widely assumed linear relationship between plasticity at different functional levels in the brain, with lower-level plasticity being understood as a prerequisite for higher-level plasticity, as we have seen, for example, in concepts of cognitive plasticity and the Hebbian notion of synaptic plasticity and learning (see Chap. 2). As is common in biology, where multiple solutions for a functional affordance are developed and prevailing, the brain exhibits both types of plasticity in parallel, at different levels, and in different functional contexts. However, this makes it important to account for different levels of plasticity, the specific ways in which plasticity is realized at each level, and how these interact in relation to specific functions. For example, at the individual level, plasticity refers to variability within unique and nonreversible developmental pathways, whereas at the species level, plasticity of specific functions is represented as a broad range of possible individual pathways and outcomes. This difference is most obvious in rare cases of unconventional brain anatomy that lack large parts of the brain (Feuillet et al., 2007; Tuckute et al., 2022; Yu et al., 2015). These cases show that psychological function and well-being do not depend on a specific spatial distribution of the brain matter in the skull, indicating a tremendous amount of anatomical plasticity in the brain. According to one of these case studies, within the skull of a French white-collar worker, all brain matter was squeezed to the outer rim and a fluid-filled bubble was positioned were usually the brain sits (see Fig. 3.1, Feuillet et al., 2007).

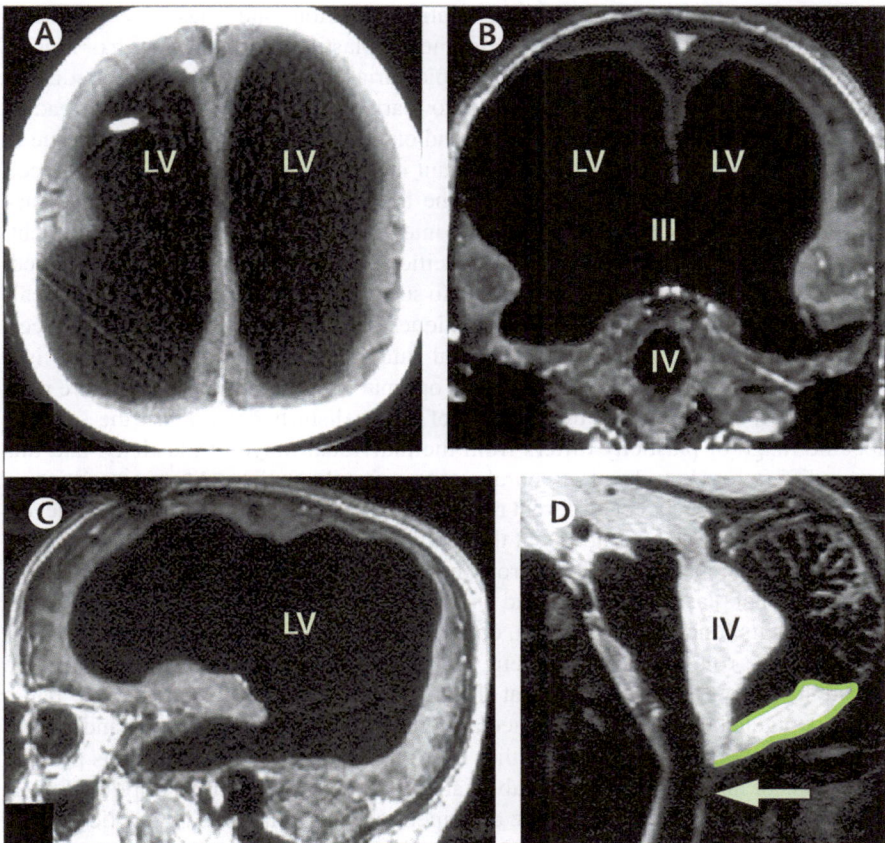

Fig. 3.1 Brain of a white-collar worker. The picture shows a massive ventricular enlargement in a patient with normal social functioning. (**a**) Computer tomography. (**b, c**) T1-weighted magnetic resonance imaging (MRI), with gadolinium contrast; (**d**) T2-weighted MRI. LV lateral ventricle. III third ventricle. IV fourth ventricle. Arrow = Magendie's foramen. The posterior fossa cyst is outlined in (**d**). (Reproduced with permission of Taylor and Francis Informa UK Ltd. through PLSclear)

This unconventional brain structure was incidentally diagnosed during a routine magnetic resonance tomography (MRT). The person did not exhibit any noticeable neurodevelopmental symptoms during his lifetime (Feuillet et al., 2007). As such rare cases show, at the species level, specialization of neural networks and brain areas, which occurs during ontogenetic development, seems to be independent from the anatomic structure to an astonishing degree. The plasticity range of neural function as it relates to the anatomic structure of brain tissue even becomes more striking if we compare it to the development of, for example, our limbs. Our capacity to walk or grasp depends strongly on the way our arms, hands, legs, and feet are positioned to our body. Here, the spatial variation and plasticity are more constrained,[2] while we see larger developmental plasticity for bone length or muscle strength, which results in higher phenotypic variability for these characteristics. In contrast, our capacity to feel emotions, to learn how to speak, or even to read or perform basic calculations does not depend on a specific position of brain tissue in the skull. What seems to be more important is the critical amount of available cell mass at a certain developmental stage, the bodily prerequisites for the necessary sensory and motor activity, and adequate interactions with our social environment. However, once a neural network for a specific psychological function is developed, it remains related to this function, at least to such a degree that we can reliably localize it across different measurement techniques. We can often even identify its cellular basis, and due to the developmental timeline of the brain tissue, there is a limited range of typical and energetically optimal neural architectures that are realized in most individuals in the absence of larger disturbances. Thus, the species-specific range of plasticity differs from the individual range of plasticity, and this range differs for every level of neural and functional plasticity. Moreover, it seems that this interplay between different ranges of plasticity at different functional levels determines functional variability at the individual and species levels.

From the perspective of the neuron, the question arises as to how the anatomical and functional plasticity of the mind and brain is represented at the single-cell level and affects its function. In addition, continuing the discussion from the last chapter regarding the role of neuroepigenetic mechanisms in single-neuron plasticity, the question is how this is reflected at the neuroepigenetic level. On the one hand, (neuro)epigenetic plasticity is defined as higher likelihood of epigenetic and related functional modifications. On the other hand, neuroepigenetic mechanisms provide the means of an energy efficient stabilization of cell states and transcriptional feedback cycles which contribute to the alignment of homeostatic and synaptic plasticity. Is neural plasticity at the neuroepigenetic level represented as flexibility and enhanced environmental sensitivity of epigenetic mechanisms, or as energy-efficient canalization and stabilization of specific transcription cycles enabling synaptic or other forms of plasticity at higher ordered functional levels? In what sense does the epigenome represent an epigenetic trace of environmental signals recorded during a neuron's life course? Also, if this is the case, do epigenetic markers provide an additional layer of information to characterize neural function, and how does our neuron theory need to change to account for and make use of this layer of epigenetic information?

---

[2] Embodied cognition approaches in robotics use this fact to reduce the computational needs for smooth movements by mimicking anatomic structures of living organisms when building robots (Pfeifer et al., 2007a, b).

Increased Stabilization as Process of Specialization: Waddington's Chreode

When considering concepts of stabilized developmental processes in combination with genetics and epigenetics, one name comes to mind: Conrad Hal Waddington. Waddington is well known as the scientist who coined the term "epigenetics" (Waddington, 1942). At his time, he was an important protagonist in the early days of twentieth-century Theoretical Biology and an advocate for the use of complex systems theory to model biological processes. He specifically attempted to reconcile embryology with population genetics and evolutionary theory (Waddington, 1940). Waddington developed a pathway model of development applicable to biological differentiation processes ranging from cellular to phylogenetic development and illustrated it with the picture of the "epigenetic landscape" (Waddington, 1942, 1957, Fig. 3.2; for a detailed history of the origins of this graphic, see Parnes, 2007). The notion of increased stabilization, representing functional specification, is one of the underlying principles of this model.

Waddington further characterized the developmental pathways crossing the epigenetic landscape as "chreodes" (1957, p. 32). These chreodes are pathways of canalized development that are relatively stable and, at the same time, developing and thus in constant flux (Waddington, 1957, for a detailed discussion and historic

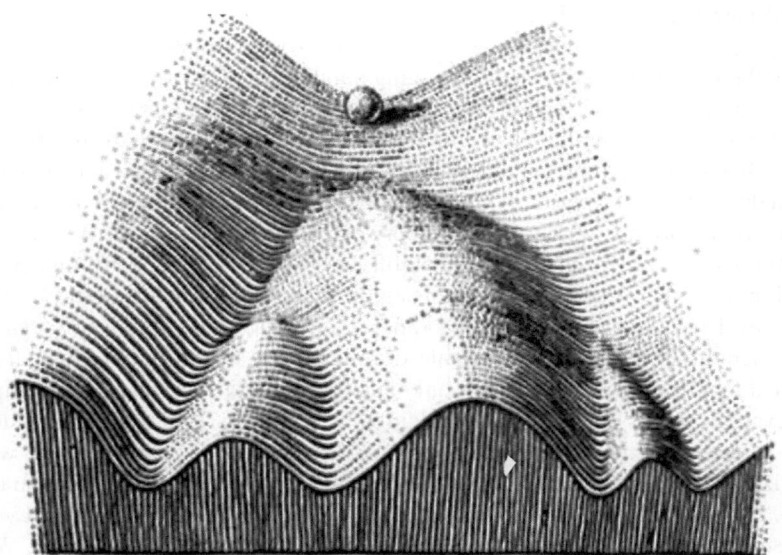

Fig. 3.2 Waddington's epigenetic landscape. The path followed by the ball represents the chreode of a cell, organ, organism, or other developmental unit. Valleys left and right from the main path depict alternative developmental trajectories (e.g., leading to different cell types, organisms, species). (Published first by C. H. Waddington in *The Strategy of the Genes*, 1957. Reproduced with permission of Taylor and Francis Informa UK Ltd. through PLSclear)

contextualization of the chreode concept, see Lux, 2016). Waddington emphasized this tension between stability and change with the word "homeorhesis," which, in reference to homeostasis, highlights the relative stability of ongoing developmental processes in motion (1957, p. 32). Chreodes represent models of this moving equilibrium for certain developing cells, organs, an organism, a population, or a species. The idea is one of a stable but moving system, canalized by the two main characteristics of the chreode: (1) the profile of the chreode, which constitutes how and in what frequency phases of slow and fast changes of the system alternate, and (2) the tendency of the system to recover to the original pathway after disturbances (Waddington, 1957, pp. 32–34). Thus, Waddington depicted development as a pathway from less specialized states with undifferentiated potentiality and a variety of outcomes to increasingly specialized functions and reduced outcome possibilities. In his view, every development has an endpoint at which the cell or organ reaches a steady state, and homeorhesis transforms into homeostasis. This marks a clear difference from other accounts of developmental systems theory, for which development is ongoing throughout the lifespan of an organism or organism–environment complex (see, for example, Bateson & Gluckman, 2012; Gilbert et al., 2015; Oyama, 2000).

That development is about reducing variability and increasing specialization, is originally an embryological argument. Waddington's view of development and Prochiantz's understanding of brain plasticity are fully in line with this embryological notion of development (Lux & Richter, 2014; Rees, 2016). As previously discussed (see Chap. 2), this contrasts with newer understandings of synaptic and non-synaptic plasticity, which address the phenotypic variability of single neurons during their cell cycle, such as the changing numbers of dendritic spines or changes in pre- and post-synaptic sensibility. Once established, these function as reversible feedback loops that fine-tune a neuron's activation state according to specific affordances at a particular time point in its activation history. How can we integrate these two understandings of plasticity into our concept of psychobiological development and, more importantly, in our neuron theory? Here, another aspect of Waddington's concept of canalized development is helpful. In contrast to the idea that higher-level functions of complex systems emerge from the interactions of parts of the system at lower levels (*emergence theory*), Waddington introduced the notion of "organization." According to him, the available developmental conditions in their diverse physical manifestations provide not only guard rails but also a broad draft template for developing cells or organisms (Waddington, 1940, 1957). In biological systems, there is always some form, structure, or gradient that guides development, which canalizes it or at least helps stabilizing the pathways along which canalization takes place. Prochiantz refers to a similar idea, which he calls "body image" and which represents the implicit blueprint of the organism (Prochiantz, 2012). While Prochiantz locates the body image within the genes, Waddington considers organization as the result of complex interactions between genes and the environment that are shaped by the organism's biography (Lux, 2016). The DNA is embedded within a living cell, which provides the structure and resources for gene expression, and this cell is embedded in serum or tissue, which is embedded within an organism or egg, etc. Thus, for Waddington, the old chicken-or-egg dilemma is not a dilemma but a necessary interdependence with one being part of the canalized pathway of the other, providing not only the genes but also huge parts of the developmental environment of the other and vice versa.

Waddington developed his notion of organization in the early 1940s based on Spemann's organizer concept and the embryological concept of induction (Waddington, 1940; see Lux, 2016). This line of thought was slightly lost in his later work, which focused more on gene-gene interaction effects, as, for example, in his pictogram of the epigenetic landscape, which focused more on gene-gene interaction effects, as, for example, in his pictogram of the epigenetic landscape. Still, it is the developmental history of the single cell and the surrounding tissue which represented for Waddington the ultimate "epigenotype" in his sense of the term.[3] He associated variability or plasticity with environmental disturbances that force the organ or organism to change its developmental trajectory against its original tendency.[4] Overall, Waddington's chreodes demonstrate how a dynamic and systemic model of development does not necessarily imply that processes and characteristics are understood as more plastic and flexible. Instead, it provides a dynamic and systemic perspective on how plasticity, specificity, and stability develop and are maintained throughout a cell's or an organism's life span.

Thus, when attempting to integrate the different notions of plasticity used to describe the functional adaptability of a single neuron, Waddington's chreode provides a level of description of the overall formative developmental pathway. This pathway canalizes the range of possible changes that can occur within the feedback loops underlying short-term synaptic or homeostatic plasticity but also stabilizes these cycles over time. In addition, for an energy-efficient realization of both types of plasticity, the energetic costs for changes within these feedback loops need to be much lower than those for changes of the whole chreode, whereas the energetic

---

[3] It is important to note that Waddington uses a developmental epigenetic concept, which differs substantially from the molecular epigenetic concept used in the study of DNA methylation and histone modifications and includes all kinds of epistatic effects of gene-gene and gene-environment interactions during development (Lux & Richter, 2014). Interestingly, the current molecular interpretation of this developmental epigenetic concept accounts for molecular epigenetic mechanisms only in regard to their function in development and specifically cell differentiation, leaving aside other functional contexts of epigenetic mechanisms such as synaptic or structural-genomic epigenetic mechanisms (Lux, 2013). Prochiantz is following such an interpretation of epigenetics as do Lerner and Overton (Lerner & Overton, 2017; Prochiantz, 2012).

[4] This gene-environment interaction model also provided the theoretical framework for his heat shock experiments in *Drosophila melanogaster*. In these experiments, he applied extreme heat to the flies' eggs and reported disturbances in the wing structures of the developing flies which, when applied repeatedly in several generations, were at one point reproduced in subsequent generations without the heat shock stimulus (Waddington, 1953). He interpreted his observations as proof of genetic assimilation through environmentally induced changes of the genetic basis of the wing-forming chreode (Waddington, 1953). This genetic stabilization of the adapted chreode, which provides the new organizational template for wing formation, shows how the available range of developmental plasticity changes under the impact of an environmental stimulus, here, the heat shock.

costs for chreode stabilization need to be much lower than those for changes of the chreode. Here, the question arises as to which biological or functional levels this canalization of plasticity is realized and detectable: the single-cell level, the immediate brain tissue context, the level of neural network connections, or a functional behavioral level, etc.

Change and Stabilization Across Levels of Plasticity

In the previous chapter, I discussed how plasticity is understood at some of these levels. Each of these concepts implicitly assumes that plasticity at a specific level impacts plasticity at other functional levels. Within Hebb's notion of plasticity, for example, according to which behavioral plasticity in the form of learning corresponds to changes in synaptic plasticity, it is assumed that flexibility and change at one level are reflected in flexibility and change at every lower level. Vice versa, behavioral stability is assumed to correspond to stability at the neural-network level and synaptic connections. However, this does not always need to be the case. For some functions, flexibility at a higher level will also depend on stability at lower levels, whereas functional stability at higher levels depends on flexibility at lower levels or even a mixture of flexibility and stability, as in the example of male zebra finches and the changing neural correlates of their song-singing behavior. Here, Waddington's notion of canalization helps us grasp such complex combinations of stability and plasticity.

It also provides a shift of perspective on specific cases of neuroplasticity, for example, the reorganization of the somatosensory cortex due to sensory loss (Singh et al., 2018). Research investigating sensory loss has repeatedly shown that specialized neural connections in the somatosensory cortex are less fixed than previously thought. Instead, the involved brain areas can be recruited by a different sensory input, be it of the same sense (e.g., the other eye) or even of a different sense, when the primary connections lose functional relevance. The common interpretation of this phenomenon is that the underlying neural networks and the neurons forming these neural networks exhibit an overseen amount of plasticity, in the sense of changeability, to make this happen. This interpretation supports the notion of a permanent flexible brain, for which the amount of change or range of plasticity is undetermined, which is in contrast to earlier embryological models of neural, brain, and psychobiological development.

The story changes dramatically when we discuss this flexibility as lost canalization, or in Waddington's terms, as a destabilized chreode. Instead of focusing on how the remaining neural projections expand to the neighboring cells and how, for example, active visual pathways win over inactive visual pathways, an alternative interpretation would be that the observed changes result from the neurons losing their primary canalized input. This lost canalization then leads to reduced functional specialization, which results in less filtering of incoming information from connected neurons and glia cells. Due to the reduced filter, less of the input stream is categorized as noise, opening up the neuron to react to a broader range of signaling information. To some extent, these neurons are reset, and one can even characterize them as "rejuvenated" to a certain degree, when we think of functional specialization of neurons as progression along their life span chreode. However, this resetting cannot place the neuron into an actual earlier state, but only into a state of similar openness and sensitivity to newly canalizing input. The cellular history will always be present to some degree in the material substances of the cell and its surroundings. In this more open state, however, the neuron is then ready and awaiting to be newly integrated into another network, which recruits it for a different function (e.g., the other eye) and forces it into a new form of specialization and stage of stability. Thus, a new and different canalization would replace the lost canalization and the chreode would reach a new state of relative stability.

This understanding of plasticity not only reconciles the embryological and the dynamic systems perspectives. It also allows us to integrate homeostatic and synaptic plasticity within a developmental pathway model, as it necessarily builds on both, with homeostatic plasticity representing the degree of openness and activation readiness and synaptic plasticity representing the experience-dependent sensitivity to specific signaling input and the direction of canalization. It also fits with the observed reduction of plasticity at the single-cell level and broader tissue level over the lifespan, and it helps to explain the role of slightly increased rates of adult neurogenesis in neural structures which contribute to psychological functions that depend on higher numbers of still very plastic neurons. In line with this, one question for future research also needs to be whether this is always compensatory (Singh et al., 2018), as suggested by the notion of one eye taking over the projection side for both eyes, or whether some forms of cognitive decline are related to the inability to keep canalization in place.

At the level of the single neuron, this change of perspective changes our picture of neural activity twofold. First, it opens the possibility that there is still a lot of variability and fuzziness in the activity of a single neuron in these networks and that this fuzziness is simply canalized by the network to the degree that the function is preserved. Second, the specific cell characteristics that allow a specialized and stable contribution of a single neuron to these networks are likely to be impacted by and result from the network input as well as the input from the surrounding cell environment. Thus, from this perspective, stability in neural activity is the product of constant *stabilization processes*, a systemic state to which the entire surrounding tissue contributes. However, it is nearly impossible to detect these stabilization processes in vivo. At the level of neural activity, we can only detect the current activity state unless we record the restructuring process in a neural network in real time. This is extremely work- and cost-intensive and not feasible in most experimental settings, let alone in clinical studies. Here, epigenetic data provide an additional layer of information that allows partial tracing of previous neural cell states if they manifest as epigenetic modifications. In particular, DNA methylation patterns have biophysical properties that function as a form of biography in this sense.

In terms of the proposed functional differentiation of epigenetic mechanisms in structural-genomic, synaptic, and developmental, functional reorientation processes at the single-neuron level are probably, at the first stage, reflected as changes in synaptic epigenetic mechanisms. We see such changes in epigenetic patterns, for example, in neurological patients after a stroke or otherwise caused brain lesions (Abu Hamdeh et al., 2021; Cullell et al., 2022; Endres et al., 2001; Wong & Langley, 2016; for a recent review, see Choi et al., 2022). In addition, developmental periods can be traced at the epigenetic level, during which changes are more likely to occur or have long-term impact (Dunn et al., 2019; Nagy & Turecki, 2012; Price et al., 2019; Provençal et al., 2020). Once developmental epigenetic mechanisms are well established for a specific function, we can infer the likelihood of specific lifetime events during these periods from the measured epigenetic markers later in life, at a time when the actual input is no longer present but the related epigenetic traces are still detectable. Taking the song-singing behavior of the zebra finch as an example, this understanding assumes that the underlying neurons need to incorporate the functional specialization during the respective developmental period, for example, when the finches acquire the song through learning. This stabilization at the single-neuron level is a prerequisite for the neuron's later contribution to the song-singing execution at specific time points. The observed plasticity at the network level, with varying groups of participating neurons over time, strongly depends on this persistent stabilization of the functional specialization at the single-neuron level, even during periods when a neuron is not currently participating in the behavioral execution. Epigenetic mechanisms are a perfect molecular fit for this task, as some of them are relatively stable and change only at some energetic costs, whereas others exhibit short-term flexibility. For example, developmental epigenetic mechanisms could function in a way that locks in the functional specialization of the neuron during the song-learning process, while the actual participation of a specific neuron in a specific moment or day of song singing is regulated by additional transcriptional mechanisms with the involvement of further synaptic epigenetic mechanisms. Both are likely intertwined and maintained by structural-genomic epigenetic mechanisms.

Another example of this complex interplay of stability and change across functional levels is the involvement of sensory and motor neurons in the exhibition and perception of an action, which has been discussed as "mirror neuron" phenomenon (Rizzolatti & Fogassi, 2014). Here, it was observed that the developmentally fixed specialization of the single neurons for a certain sensory and motor modality of a specific action allows them to be integrated in neuronal networks related to the exhibition of the action, as well as to the perception or imagination of the action. However, different firing rates have been reported at the single-neuron level in some of these highly specialized neurons, depending on their function at the behavioral level: motor action, perception, or imagination (Bonini et al., 2010; Fogassi et al., 2005; Pellicano et al., 2021; Rizzolatti & Sinigaglia, 2007). Thus, there is some time-accurate representation at the single-neuron level of these different functions when the neuron participates in them. In addition, for each function, specialized neurons are recruited by other neurons, which canalize the functional involvement of the specialized sensory and motor neurons to ensure their participation in the exhibition of higher cognitive processes at a specific time point. This canalization must also be present at the single-neuron level in the form of a partially functional, yet short-term, and reversible specialization. Here, epigenetic mechanisms provide potentially matching biophysical properties, contributing to both long-term and short-term differentiation. However, the exact mechanism requires further investigation.

These examples also show an important component of the plasticity–stability complex. Not only is there always some degree of specialization and non-specialization at the single neuron and at the neural network level, possibly represented by functionally different epigenetic mechanisms across different neurons in a certain brain area or neural network. Moreover, the hierarchical embeddedness of a single neuron, its recruitment by a specific neuronal network, is canalized by mechanisms at the functional level: the sensory or motor input (visual system and mirror neurons action and perception), a complex combination in the form of a behavioral pattern (zebra finch song singing), or different higher cognitive processes (mirror neurons in action and action imagination). Thus, these higher levels canalize and, at the same time, stabilize the underlying neuronal networks due to adequate feedback processes. The song-singing behavior of the zebra finch recruits appropriate neural networks. These rely on previously established—through learning—and stabilized neuronal specializations. In addition, their stabilization at a higher functional level allows for some flexibility at the lower levels; at the same time, the functional stability at higher functional levels relies on the canalization and buffering mechanisms at each of the lower levels.

Furthermore, the hierarchical relationship of the function between these different levels of observation shows a clear time component. First and foremost, this time component represents the difference between learning an action and executing a learned action, represented by the difference between establishing a neural network and its subsequent activation. However, each activation also re-establishes the network at new. On a phenomenological level, this resembles the subjective experience of riding a bike. Once we reach a basic level of general bike-riding ability, we need to pay less attention to the movements and profit from some form of automatized coordination. Nevertheless, we can always improve the coordination of our movements and the speed at which we exhibit the sequence of movements necessary to ride a bike, which is what professional athletes constantly train. In addition, every bike and every bike ride are slightly different, so we adjust our movements accordingly. This may be why getting back on a bike after years of not riding a bike often feels awkward and familiar at the same time. At the neural activity level, learning requires stabilization. However, to identify the different mechanisms that contribute to the plasticity-stability complex, we need a model of psychobiological development that enables us to analytically differentiate between the different levels at which we record change and stability.

Gottlieb's Bidirectional Model of Psychobiological Development and Beyond

Gilbert Gottlieb's (2007) model of psychobiological development represents such a hierarchical bidirectional model. Further specifications of the model include a more precise integration of the social-cultural context of human psychological development (Valsiner, 2007) and the addition of a level of epigenetic activity, as well as a hormonal context layer (see Fig. 3.3; Lux, 2013). According to Gottlieb's model, bidirectional interactions at lower levels support the development of higher-order levels, which then take over the canalization of lower levels. However, higher levels or functions are not understood as emerging from the interaction of parts at lower levels. Instead, Gottlieb emphasized that higher-order levels are established through *coactive* development, which includes organic, tissue-related, material, and functional reorganization (Gottlieb, 2007). Even for higher psychological functions, he points out that there is always some sort of canalization in place due to the environmental and bodily conditions of psychobiological development. This includes the physical (temperature, nutrients, oxygen, toxins) environment and the hormone system, as well as bone structures such as the skull (Gottlieb, 2007; Lux, 2013). In addition, social-cultural interactions, which are a vital part of human psychobiological development, also provide some form of stabilization and canalization (Valsiner, 2007).

The model allows us to analytically differentiate between levels of observation. Each level accounts for a different epistemological entity, such as a motor action

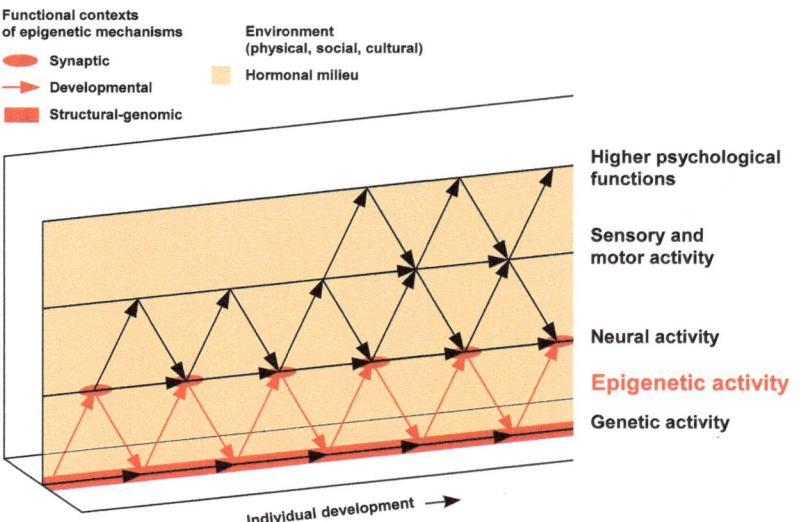

Fig. 3.3 Modified model of psychobiological development with epigenetic mechanisms. Three functional contexts of epigenetic mechanisms are distinguished: structural-genomic, developmental, and synaptic. (Figure adapted from Lux, 2013, CC BY-NC 4.0 Deed)

versus a neural activation pattern, as well as the methods of description and data collection used to register characteristics at each of these levels. Each level also represents a biomaterial and psychophysiological entity that provides certain functional characteristics to the process of psychobiological development. Thus, their interdependence is also a product of their material embeddedness. For the neural activity level, this means, for example, that it is embedded within motor and sensory activities and higher psychological functions, and at the same time, it is stabilized by and stabilizes the lower genetic and epigenetic activity levels. In this sense, each level also represents a level of transmission and translation. Neural activity mediates between the single-neuron level and neural networks, with neuroepigenetic mechanisms being one part of how this embeddedness is realized at the molecular single-cell level and translated into genetic activity and vice versa. Every development at a higher level is restructuring lower levels to some extent and may restructure them even further over time. In addition, development at higher levels depends on that at lower levels. Gottlieb strongly emphasized this coactive relationship between the levels and, specifically, its bidirectionality (Gottlieb, 2007). He primarily discusses bidirectionality within the context of developmental change, focusing on the unfolding of new structures and functions. However, it is also central to the processes of stabilization and canalization, which aim to preserve function. Finally, the developmental model does not allow for reversibility but, to some extent, for reorganization at each level across the lifespan.

However, when using these level distinctions, we must bear in mind that this hierarchical separation is only analytical. In the individual, all levels act together at the same time, and this makes it difficult to not confuse them or reduce them to each other when we measure activity at individual levels in real time. Overall, but also at each level separately, we find tremendous biological fuzziness, redundancy, and indeterminacy. These result partly from combining different and multiple measurements when registering activity at each level and partly represent key characteristics of biological processes in general. In an effort to relate the different analytical levels to each other, we need theoretical concepts and models to bridge them in a way that fits the empirical observations.

Self-Stabilization and Interactive Specialization

Several propositions have been made that try to capture the observed complexity at different levels within neural networks and how they relate to single-neuron activity as well as behavior reaching from connectionist approaches to neurophenomenology. One attempt that explicitly uses a hierarchical model together with a developmental perspective is a computation-based approach to developmental systems neuroscience based on dynamic field theory (Schöner & Spencer, 2015). The aim of this approach is to provide a mathematical description for models that bridge the neural activity level of small neuron populations with the sensory and motor activity level, as well as some higher cognitive processes such as working memory and decision-making. This approach assumes that the fuzzy multiplicity of neural activation patterns, resulting from a myriad of potential signals and random neural activity, is stabilized over time, and that this stabilization takes place at the level of small populations of neurons based on a privileged sensory input. These stabilized activity patterns are then understood to represent the neural activity underlying the mental states related to a specific perception, motor activity, or cognitive process (Schöner, 2014, 2020).

The stabilization process itself is modeled as an attractor of the dynamic system within this dynamic field theory approach (Schöner et al., 2015; Schöner & Schutte, 2015). It is initiated by the input at the higher-level sensory or motor activity, after which the system enters a phase of self-stabilization during which the neural activity pattern is still present in the brain independent of the sensory input and the immediate microscopic activity of single neurons activated by it. These prolonged activity patterns at the level of small neural populations are then interpreted as representing the neural trace underlying mental states and behavior: "Cognition and behavior emerge from such networks simply in the sense that the time courses of macroscopic activation patterns that are generated in these networks constitute the time courses of mental states, which are coupled to sensory information and ultimately drive muscles to bring about bodily movement" (Schöner, 2020, p. 1268). Accordingly, within this approach, the level of summed, self-stabilizing neural activity cycles is seen as a privileged level of description in cognitive neuroscience (Schöner, 2014, 2020). In addition, by understanding the attractor of a specific sensory input or motor activity as being established over time and canalized by interaction with the environment, the model can account for experience-dependent specialization of brain areas and networks during the course of brain and psychobiological development, as observed in infant cognition and motor learning (Aerdker et al., 2022; Dineva & Schöner, 2018). However, in contrast to Gottlieb's model, Schöner et al. clearly emphasize that the modeled neural activity, although attempting to capture the spatial and temporal dynamics of neural activation, does not represent the activity of single neurons or biological processes at cellular or molecular levels that enable them (Schneegans et al., 2015; Schöner et al., 2015; Schöner & Schutte, 2015).

This marks a clear difference to the approach put forward by Marc H. Johnson, which directly builds on Gottlieb's probabilistic epigenesis. He proposes understanding this process of stabilization and canalization over time as "interactive specialization" *at all levels*, ranging from the single-cell level to brain areas and larger neural networks (Johnson, 2011). Based on this notion, Johnson outlines a progressive specialization of neural networks and domains throughout brain development, which is accompanied by dynamic processes of reorganization within smaller and more global neural networks serving specialization at higher levels and domains (Johnson, 2000, 2011; Johnson & Haan, 2015). At the single-cell level, these reorganization processes are represented by the integration of newly developing neurons or, at a later stage, the restructuring of existing neural connections based on the principles of Hebbian synaptic plasticity. The concept of interactive specialization relies on several assumptions: "(1) cortical areas are inextricably linked through dense patterns of interconnections that contribute to coordinated sequences of development, (2) probabilistic epigenesis gives a vital role to intrinsically and extrinsically generated activity in sculpting anatomical development, (3) combinations of cortical regions may support similar or identical behaviors in different ways during the course of development, and (4) plasticity is the inherent state of an unspecialized neural system"(Johnson, 2020, p. 278). According to Johnson, early in development, brain areas and groups of neurons as well as individual neurons within local populations still exhibit a wide range of plasticity. Within his interactive specialization approach, he defines plasticity as "the state during which a region's function is not yet fully specialized. That is, there is still remaining scope for developing more finely tuned responses, and a reduction in plasticity is simply a byproduct of the process of development" (Johnson, 2020, p. 278). This echoes the notion of development constraining possible pathways and alternative endpoints of the developmental process, as outlined within Waddington's epigenetic landscape and his chreode model.[5]

This approach hypothesizes that brain areas and local neural networks develop asynchronously, being integrated step by step in the representation of specific functions within the process of their development. Broad and unspecific local representations develop into more pronounced and fine-tuned functional representations at the network level, with smaller and more specific local networks, which, on the other hand, become integrated into larger global networks as higher cognitive processes evolve. This also implies that local representations, as well as the form of integration at the large-scale level, change with development, specifically between childhood and adulthood. Therefore, interactive specialization provides a middle ground between the notion of local representation within specialized regions and networks and a large-scale connectome approach. Furthermore, interactive specialization corresponds with the view that atypical early experiences can have long-term effects on development because they affect the process of functional specializations of networks and thus shape the course of development. Thus, the approach provides an ecological account of the development of functional specialization at the level of individual regions and small functional networks related to specific cognitive tasks.

Both approaches, developmental systems neuroscience and interactive specialization, emphasize in their models the stabilization effect of hierarchical canalization as well as the time component of this canalization effect. Functional stabilization is then achieved when the changes or the state induced by the signal from the higher level is still present in the absence of the signal. Within the framework of developmental systems neuroscience, this is modeled as self-stabilization process at the level of small neural networks, and the contribution of single neurons to this effect is not measured, while, according to the notion of interactive specialization, functional stabilization is realized by bidirectional fine-tuning mechanisms that include

---

[5] Johnson directly refers to Waddington's epigenetic landscape in his introductory text book *Developmental Cognitive Neuroscience: An Introduction* in the 4th edition coauthored together with Michelle de Haan (Johnson & Haan, 2015).

cell differentiation and functional specialization at the single-neuron level. When raising the question of how this functional specialization is realized at the single-neuron level, we, again, can turn to neuroepigenetic mechanisms and their biophysical properties, providing a relatively stable and environmentally sensitive mechanism of functional specialization at the molecular and chromatin level, affecting gene expression and protein synthesis. In this context, the study of epigenetic modifications in brain tissue and single brain cells provides new data to trace plasticity, stability, and canalization at the molecular level.

Neuroepigenetics: A New Level to Register Differences in Plasticity

At the molecular level, the question arises as to why those mechanisms classified as neuroepigenetics should receive more attention than other molecular representations of cellular functions. One could argue that the proteome or transcriptome is a much more precise representation of what is occurring in a cell. The detection of neurotransmitters and their molecular counterplayers, as well as related transcription factors, contributes tremendously to the understanding of how a single neuron works. However, one issue with these profiles is their relative instability. The proteome and transcriptome of a cell change very quickly. Both represent the present cell state, at least in the main part. They are ideal for detecting short-term molecular responses to specific neural activation. The genome of a neuron, on the other hand, is too stable to sensibly register changes throughout the lifetime of a neuron. The epigenetic mechanisms provide an intermediate level. They show relative stability and sensitivity to neural activity and other forms of influence from different cell signaling systems and environmental conditions. Epigenetic mechanisms allow us to register both short-term (synaptic epigenetic mechanisms) and long-term (developmental epigenetic mechanisms) changes at the molecular level, with potential effects on neural function. Again, this differentiation between proteome, transcriptome, epigenome, and genome is only analytical. Some parts of the proteome or transcriptome, some miRNA fragments, as well as DNMTs or HATs, clearly participate in epigenetic mechanisms, and the genome strongly influences which epigenetic modifications are present and possible (see Chap. 5). The question has rightly been debated as to whether it makes sense at all to classify epigenetic mechanisms independently of transcription factor activity (see Horsthemke, 2022, and Chap. 2). As long as we are only interested in the molecular underpinnings of the present neural activity, or "synaptic epigenetic mechanisms," it would be sufficient to describe them as feedback loops and molecular cell cycles of transcription factors leading to changes in neurotransmitter availability and so forth. However, as soon as we assume that a previous impact influences these feedback loops in a cell-specific but enduring manner, we need a level of detection that allows us to differentiate between changes at different timeframes. Epigenetic mechanisms enable the detection of modifications at different timeframes (see Chap. 5). They represent a specific level to register parts of the molecular underpinnings of the plasticity–stability complex involved in neural function.

First examples of neuroepigenetic studies show the potential of this additional level of data collection. Wada et al. (2013) studied the neural correlates of differences in song-singing behavior between a wild type and a domesticated variant of a songbird, the Bengalese finch. The domesticated birds exhibited a higher variability in song-singing behavior. Studies of the underlying neural networks showed that both strains use the same neural pathways during song singing. Thus, behavioral differences were not detectable at the neural network level. Instead, the study found differences in DNA methylation patterns at the androgen receptor gene in GABAergic neurons of the basal ganglia nucleus Area X between the two strains. Area X is a brain area specific to songbirds that strongly contributes to song-singing behavior. This example shows that behavioral plasticity in this species is stabilized at the molecular level within a certain type of neuron in the underlying neural networks. These epigenetic modifications contribute to the stabilization and canalization of species-specific behavior, leading to different phenotypes in the two strains at the population level. The modification represents a developmental epigenetic mechanism that impacts inner species-specific differentiation and could even have a long-term impact on species development within an evolutionary timeframe, as discussed by the authors of the study (Wada et al., 2013).

Another neuroepigenetic study demonstrated the stabilizing role of epigenetic mechanisms in neural networks, and the consequences when this stabilization is no longer in place. Rusconi et al. (2016) reported results from a mice model according to which chronic stress together with the loss of epigenetic control over the transcription of immediate early genes (IEGs) in the hippocampus leads to vulnerability and corruption of corticolimbic circuits and was associated with depression-like behavior. They concluded that this is because the loss of epigenetic control leads to exaggerated plasticity in the hippocampi of these mice. In another study, they found that decreased IEGs (*Arc* and *Erg1*) transcription in the ventral hippocampus was related to stress resilience in these mice (Bagot et al., 2015). Furthermore, an optogenetic reduction of neuroplasticity via the induction of LTD in the circuit connecting the ventral hippocampus with the nucleus accumbens reduced depression-like behavior induced by chronic stress in these stress-vulnerable mice (Bagot et al., 2015).

These examples show that epigenetic markers provide an additional level of data collection, which allows tracing and characterization of molecular correlates of neural function. Moreover, if we are able to specify these epigenetic markers further, they can provide information on the molecular context of neural plasticity, including mechanisms involved in the canalization of neural function along different timeframes from short-term synaptic to lifelong developmental and evolutionary timeframes. In contrast to the notion that dynamic epigenetic mechanisms make the neuron more plastic and flexible—a picture that originates in the assumption that epigenetics stands in opposition to genetic determinism—these examples also show that, in many cases, epigenetic modifications stabilize or canalize plasticity. To the most part, epigenetic modifications provide an induced cell memory—not to be confused with memory as psychological function[!], which is established at high energy costs. However, once established, it is also erased only at some energetic cost, thereby providing an ideal molecular mechanism for the specialization process a neuron goes through during its single-cell history. In addition, these epigenetic modifications still need to be maintained to some extent, and the establishment and maintenance of epigenetic modifications highly depend on the inner and outer cellular context. Even if we understand the maintenance of epigenetic modifications similar to other cellular characteristics as stabilized by catalytic self-sustaining loops at the individual cell level, in neurons, these catalytic loops are influenced and often initiated by neural function, which is directed by the higher functional levels of sensory and motor activity and higher psychological functions. Furthermore, as Gottlieb's model and its extensions remind us (Gottlieb, 2007; Lux, 2013; Valsiner, 2007), sensory and motor activity and higher psychological functions are embedded within the biomaterial entities of the living body and a social-cultural context. Thus, if we would like to understand the plasticity and stability of epigenetic modifications in neurons, we need to ask how they are set up and, maybe even more important, how they are kept in place. For this purpose, we must turn to the neuron in its context.

References

Abu Hamdeh, S., Ciuculete, D.-M., Sarkisyan, D., Bakalkin, G., Ingelsson, M., Schiöth, H. B., & Marklund, N. (2021). Differential DNA methylation of the genes for amyloid precursor protein, Tau, and neurofilaments in human traumatic brain injury. *Journal of Neurotrauma, 38*(12), 1679–1688. https://doi.org/10.1089/neu.2020.7283

Aerdker, S., Feng, J., & Schöner, G. (2022). Habituation and dishabituation in motor behavior: Experiment and neural dynamic model. *Frontiers in Psychology, 13*, 717669. https://www.frontiersin.org/articles/10.3389/fpsyg.2022.717669

Bagot, R. C., Parise, E. M., Peña, C. J., Zhang, H.-X., Maze, I., Chaudhury, D., Persaud, B., Cachope, R., Bolaños-Guzmán, C. A., Cheer, J., Deisseroth, K., Han, M.-H., & Nestler, E. J. (2015). Ventral hippocampal afferents to the nucleus accumbens regulate susceptibility to depression. *Nature Communications, 6*, 7062. https://doi.org/10.1038/ncomms8062

Bateson, P., & Gluckman, P. (2012). Plasticity and robustness in development and evolution. *International Journal of Epidemiology, 41*(1), 219–223. https://doi.org/10.1093/ije/dyr240

Bonini, L., Rozzi, S., Serventi, F. U., Simone, L., Ferrari, P. F., & Fogassi, L. (2010). Ventral premotor and inferior parietal cortices make distinct contribution to action organization and intention understanding. *Cerebral Cortex (New York, N.Y.: 1991), 20*(6), 1372–1385. https://doi.org/10.1093/cercor/bhp200

Choi, D.-H., Choi, I.-A., & Lee, J. (2022). The role of DNA methylation in stroke recovery. *International Journal of Molecular Sciences, 23*(18), 10373. https://doi.org/10.3390/ijms231810373

Cullell, N., Soriano-Tárraga, C., Gallego-Fábrega, C., Cárcel-Márquez, J., Muiño, E., Llucià-Carol, L., Lledós, M., Esteller, M., de Moura, M. C., Montaner, J., Rosell, A., Delgado, P., Martí-Fábregas, J., Krupinski, J., Roquer, J., Jiménez-Conde, J., & Fernández-Cadenas, I. (2022). Altered methylation pattern in EXOC4 is associated with stroke outcome: An epigenome-wide association study. *Clinical Epigenetics, 14*(1), 124. https://doi.org/10.1186/s13148-022-01340-5

Dineva, E., & Schöner, G. (2018). How infants' reaches reveal principles of sensorimotor decision making. *Connection Science, 30*(1), 53–80. https://doi.org/10.1080/09540091.2017.1405382

Dunn, E. C., Soare, T. W., Zhu, Y., Simpkin, A. J., Suderman, M. J., Klengel, T., Smith, A. D. A. C., Ressler, K. J., & Relton, C. L. (2019). Sensitive periods for the effect of childhood adversity on DNA methylation: Results from a prospective, longitudinal study. *Biological Psychiatry, 85*(10), 838–849. https://doi.org/10.1016/j.biopsych.2018.12.023

Endres, M., Fan, G., Meisel, A., Dirnagl, U., & Jaenisch, R. (2001). Effects of cerebral ischemia in mice lacking DNA methyltransferase 1 in post-mitotic neurons. *Neuroreport, 12*(17), 3763–3766. https://doi.org/10.1097/00001756-200112040-00032

Feuillet, L., Dufour, H., & Pelletier, J. (2007). Brain of a white-collar worker. *The Lancet, 370*(9583), 262. https://doi.org/10.1016/S0140-6736(07)61127-1

Fogassi, L., Ferrari, P., Gesierich, B., Rozzi, S., Chersi, F., & Rizzolatti, G. (2005). Parietal lobe: From action organization to intention understanding. *Science (New York, N.Y.), 308*(5722), 662–667. https://doi.org/10.1126/science.1106138

Gilbert, S. F., Bosch, T. C. G., & Ledón-Rettig, C. (2015). Eco-Evo-Devo: Developmental symbiosis and developmental plasticity as evolutionary agents. *Nature Reviews. Genetics, 16*(10), 611–622. https://doi.org/10.1038/nrg3982

Gottlieb, G. (2007). Probabilistic epigenesis. *Developmental Science, 10*(1), 1–11. https://doi.org/10.1111/j.1467-7687.2007.00556.x

Horsthemke, B. (2022). A critical appraisal of clinical epigenetics. *Clinical Epigenetics, 14*(1), 95. https://doi.org/10.1186/s13148-022-01315-6

Johnson, M. H. (2000). Functional brain development in infants: Elements of an interactive specialization framework. *Child Development, 71*(1), 75–81.

Johnson, M. H. (2011). Interactive specialization: A domain-general framework for human functional brain development? *Developmental Cognitive Neuroscience, 1*(1), 7–21. https://doi.org/10.1016/j.dcn.2010.07.003

Johnson, M. H. (2020). Chapter 13—Theories in developmental cognitive neuroscience. In J. Rubenstein, P. Rakic, B. Chen, & K. Y. Kwan (Eds.), *Neural circuit and cognitive development* (2nd ed., pp. 273–288). Academic Press. https://doi.org/10.1016/B978-0-12-814411-4.00013-5

Johnson, M. H., & de Haan, M. (2015). *Developmental Cognitive Neuroscience: An Introduction*. Wiley.

Kirkpatrick, J., Pascanu, R., Rabinowitz, N., Veness, J., Desjardins, G., Rusu, A. A., Milan, K., Quan, J., Ramalho, T., Grabska-Barwinska, A., Hassabis, D., Clopath, C., Kumaran, D., & Hadsell, R. (2017). Overcoming catastrophic forgetting in neural networks. *Proceedings of the National Academy of Sciences, 114*(13), 3521–3526. https://doi.org/10.1073/pnas.1611835114

Lerner, R. M., & Overton, W. F. (2017). Reduction to absurdity: Why epigenetics invalidates all models involving genetic reduction. *Human Development, 60*(2/3), 107–123.

Lux, V. (2013). With Gottlieb beyond Gottlieb: The role of epigenetics in psychobiological development. *International Journal of Developmental Science, 7*(2), 69–78. https://doi.org/10.3233/DEV-1300073

Lux, V. (2016). Conrad Hal Waddingtons "Chreode". In *Synergie: Kultur- und Wissensgeschichte einer Denkfigur; Trajekte* (pp. 247–263). Wilhelm Fink. http://publikationen.ub.uni-frankfurt.de/frontdoor/index/index/docId/46936

Lux, V., & Richter, J. T. (2014). Einleitung. In *Kulturen der Epigenetik: Vererbt, codiert, übertragen* (p. xiii–xxviii). De Gruyter. https://doi.org/10.1515/9783110316032.xiii

Mermillod, M., Bugaiska, A., & Bonin, P. (2013). The stability-plasticity dilemma: Investigating the continuum from catastrophic forgetting to age-limited learning effects. *Frontiers in Psychology, 4*, 504. https://www.frontiersin.org/articles/10.3389/fpsyg.2013.00504

Muñoz-Martin, I., Bianchi, S., Hashemkhani, S., Pedretti, G., Melnic, O., & Ielmini, D. (2021). A brain-inspired homeostatic neuron based on phase-change memories for efficient neuromorphic computing. *Frontiers in Neuroscience, 15*, 709053. https://www.frontiersin.org/articles/10.3389/fnins.2021.709053

Nagy, C., & Turecki, G. (2012). Sensitive periods in epigenetics: Bringing us closer to complex behavioral phenotypes. *Epigenomics, 4*(4), 445–457. https://doi.org/10.2217/epi.12.37

Oyama, S. (2000). *The ontogeny of information: Developmental systems and evolution* (2nd ed., rev.and expanded ed.). Duke University Press.

Parisi, G. I., Kemker, R., Part, J. L., Kanan, C., & Wermter, S. (2019). Continual lifelong learning with neural networks: A review. *Neural Networks, 113*, 54–71. https://doi.org/10.1016/j.neunet.2019.01.012

Parnes, O. (2007). Die Topographie der Vererbung. Epigenetische Landschaften bei Waddington und Piper. *Trajekte, 14*, 26–31.

Pellicano, A., Mingoia, G., Ritter, C., Buccino, G., & Binkofski, F. (2021). Respiratory function modulated during execution, observation, and imagination of walking via SII. *Scientific Reports, 11*(1), Article 1. https://doi.org/10.1038/s41598-021-03147-5

Pfeifer, R., Bongard, J., & Grand, S. (2007a). *How the body shapes the way we think: A new view of intelligence.* MIT Press.

Pfeifer, R., Lungarella, M., & Iida, F. (2007b). Self-organization, embodiment, and biologically inspired robotics. *Science (New York, N.Y.), 318*(5853), 1088–1093. https://doi.org/10.1126/science.1145803

Price, A. J., Collado-Torres, L., Ivanov, N. A., Xia, W., Burke, E. E., Shin, J. H., Tao, R., Ma, L., Jia, Y., Hyde, T. M., Kleinman, J. E., Weinberger, D. R., & Jaffe, A. E. (2019). Divergent neuronal DNA methylation patterns across human cortical development reveal critical periods and a unique role of CpH methylation. *Genome Biology, 20*(1), 196. https://doi.org/10.1186/s13059-019-1805-1

Prochiantz, A. (2012). *Qu'est-ce que le vivant?* Éditions du Seuil.

Provençal, N., Arloth, J., Cattaneo, A., Anacker, C., Cattane, N., Wiechmann, T., Röh, S., Ködel, M., Klengel, T., Czamara, D., Müller, N. S., Lahti, J., Null, N., Räikkönen, K., Pariante, C. M., Binder, E. B., Kajantie, E., Hämäläinen, E., Villa, P., & Laivuori, H. (2020). Glucocorticoid exposure during hippocampal neurogenesis primes future stress response by inducing changes in DNA methylation. *Proceedings of the National Academy of Sciences, 117*(38), 23280–23285. https://doi.org/10.1073/pnas.1820842116

Rees, T. (2016). *Plastic reason: An anthropology of brain science in embryogenetic terms.* University of California Press.

Rizzolatti, G., & Fogassi, L. (2014). The mirror mechanism: Recent findings and perspectives. *Philosophical Transactions of the Royal Society of London. Series B, Biological Sciences, 369*(1644), 20130420. https://doi.org/10.1098/rstb.2013.0420

Rizzolatti, G., & Sinigaglia, C. (2007). Mirror neurons and motor intentionality. *Functional Neurology, 22*(4), 205–210.

Rusconi, F., Grillo, B., Ponzoni, L., Bassani, S., Toffolo, E., Paganini, L., Mallei, A., Braida, D., Passafaro, M., Popoli, M., Sala, M., & Battaglioli, E. (2016). LSD1 modulates stress-evoked transcription of immediate early genes and emotional behavior. *Proceedings of the National Academy of Sciences of the United States of America, 113*(13), 3651–3656. https://doi.org/10.1073/pnas.1511974113

Schneegans, S., Lins, J., & Schöner, G. (2015). Embedding dynamic field theory in neurophysiology. In G. Schöner, J. Spencer, & D. Research Group (Eds.), *Dynamic thinking: A primer on dynamic field theory* (p. 0). Oxford University Press. https://doi.org/10.1093/acprof:oso/9780199300563.003.0003

Schöner, G. (2014). Dynamical systems thinking: From metaphor to neural theory. In *Handbook of developmental systems theory and methodology* (pp. 188–217). The Guilford Press.

Schöner, G. (2020). The dynamics of neural populations capture the Laws of the mind. *Topics in Cognitive Science, 12*(4), 1257–1271. https://doi.org/10.1111/tops.12453

Schöner, G., & Schutte, A. R. (2015). Dynamic field theory: Foundations. In G. Schöner, J. Spencer, & D. Research Group (Eds.), *Dynamic thinking: A primer on dynamic field theory* (p. 0). Oxford University Press. https://doi.org/10.1093/acprof:oso/9780199300563.003.0002

Schöner, G., & Spencer, J. (2015). *Dynamic thinking: A primer on dynamic field theory*. Oxford University Press.

Schöner, G., Reimann, H., & Lins, J. (2015). Neural dynamics. In G. Schöner, J. Spencer, & D. Research Group (Eds.), *Dynamic thinking: A primer on dynamic field theory* (p. 0). Oxford University Press. https://doi.org/10.1093/acprof:oso/9780199300563.003.0001

Schug, S., Benzing, F., & Steger, A. (2021). Presynaptic stochasticity improves energy efficiency and helps alleviate the stability-plasticity dilemma. *eLife, 10*, e69884. https://doi.org/10.7554/eLife.69884

Singh, A. K., Phillips, F., Merabet, L. B., & Sinha, P. (2018). Why does the cortex reorganize after sensory loss? *Trends in Cognitive Sciences, 22*(7), 569–582. https://doi.org/10.1016/j.tics.2018.04.004

Tuckute, G., Paunov, A., Kean, H., Small, H., Mineroff, Z., Blank, I., & Fedorenko, E. (2022). Frontal language areas do not emerge in the absence of temporal language areas: A case study of an individual born without a left temporal lobe. *Neuropsychologia, 169*, 108184. https://doi.org/10.1016/j.neuropsychologia.2022.108184

Valsiner, J. (2007). Gilbert Gottlieb's theory of probabilistic epigenesis: Probabilities and realities in development. *Developmental Psychobiology, 49*(8), 832–840. https://doi.org/10.1002/dev.20276

Wada, K., Hayase, S., Imai, R., Mori, C., Kobayashi, M., Liu, W., Takahasi, M., & Okanoya, K. (2013). Differential androgen receptor expression and DNA methylation state in striatum song nucleus Area X between wild and domesticated songbird strains. *European Journal of Neuroscience, 38*(4), 2600–2610. https://doi.org/10.1111/ejn.12258

Waddington, C. H. (1940). *Organisers & genes*. The University Press.

Waddington, C. H. (1942). The epigenotype. *Endeavour, 1*, 18–20.

Waddington, C. H. (1953). Genetic assimilation of an acquired character. *Evolution, 7*(2), 118–126. https://doi.org/10.2307/2405747

Waddington, C. H. (1957). *The strategy of the genes: A discussion of some aspects of theoretical biology: With an appendix by H. Kacser*. George Allen and Unwin.

Wong, V. S., & Langley, B. (2016). Epigenetic changes following traumatic brain injury and their implications for outcome, recovery and therapy. *Neuroscience Letters, 625*, 26–33. https://doi.org/10.1016/j.neulet.2016.04.009

Yu, F., Jiang, Q., Sun, X., & Zhang, R. (2015). A new case of complete primary cerebellar agenesis: Clinical and imaging findings in a living patient. *Brain, 138*(6), e353. https://doi.org/10.1093/brain/awu239

Chapter 4
Neurons in Context

In the history of tissue culture, plasticity [...] is achieved in many cases by a manipulation not just of the cell itself but also of the medium in which it lives. (Landecker, 2007, p. 11)

Cajal's silver staining method visualized the individual single neuron in brain tissue, making it possible to study its structure and function at the single-cell level. By introducing this visual contrast between the neuron and its surrounding tissue, the staining method supported a view that focused on the neuron in isolation. What followed was the idea that it is possible to single out one neuron or a network of neurons to describe the workings of the brain. As discussed above, this experimental reductionism successfully guided investigations into the specific role of subgroups of neurons, such as motor activity (Arber, 2017; Capelli et al., 2017), or the participation of single neurons in a specific function, as intensely studied in the rat's whisker system (Goldin et al., 2018; Petersen, 2007). This allowed to study the function of subparts of the neuron, most importantly the synapse (e.g., Kandel, 2000; Wang & Dudko, 2021; Watanabe, 2015). It even helped produce algorithms for machine learning approaches to cognition and artificial intelligence (for an overview, see, e.g., Kriegeskorte & Golan, 2019).

However, in real life, neurons are not isolated. They are surrounded by tissue containing thousands of other neurons in their direct neighborhood and 50 times more glia cells. These glia cells include microglia, astrocytes, oligodendrocytes, and their progenitors, NG2 glia, which exhibit different tissue preservation, cell regulation, and developmental functions. A growing number of studies have shown the involvement of glia cells in synaptic functions (Bacci et al., 1999; Krishna Temburni & Jacob, 2001; Perez-Catalan et al., 2021; Stogsdill & Eroglu, 2017). Most importantly, a single neuron is embedded in a manifold of neural networks, on average, connected with 10,000 other neurons, with some neurons, such as Purkinje cells, receiving input from up to 200,000 synapses (Dusart & Flamant, 2012). These networks are embedded in different brain areas with their specific cellular composition and different timelines of cell differentiation and development. In addition, the brain sits in the skull and is surrounded by the body. The brain is connected to various physiological systems, muscle structures, and sensory organs. Moreover, it is embedded in cerebral fluid containing hormones, neurotransmitters, and nutrients, and the brain tissue is crossed by blood vessels that deliver oxygen, glucose, and other components relevant for neural function. Every minute, approximately 700 ml blood travels through the adult brain (Zarrinkoob et al., 2015). The supporting astrocytes not only actively maintain a highly selective blood–brain barrier. They also ensure that all nutrients relevant for a neuron's function pass this barrier using a variety of molecular channels in their cell membranes (Fig. 4.1).

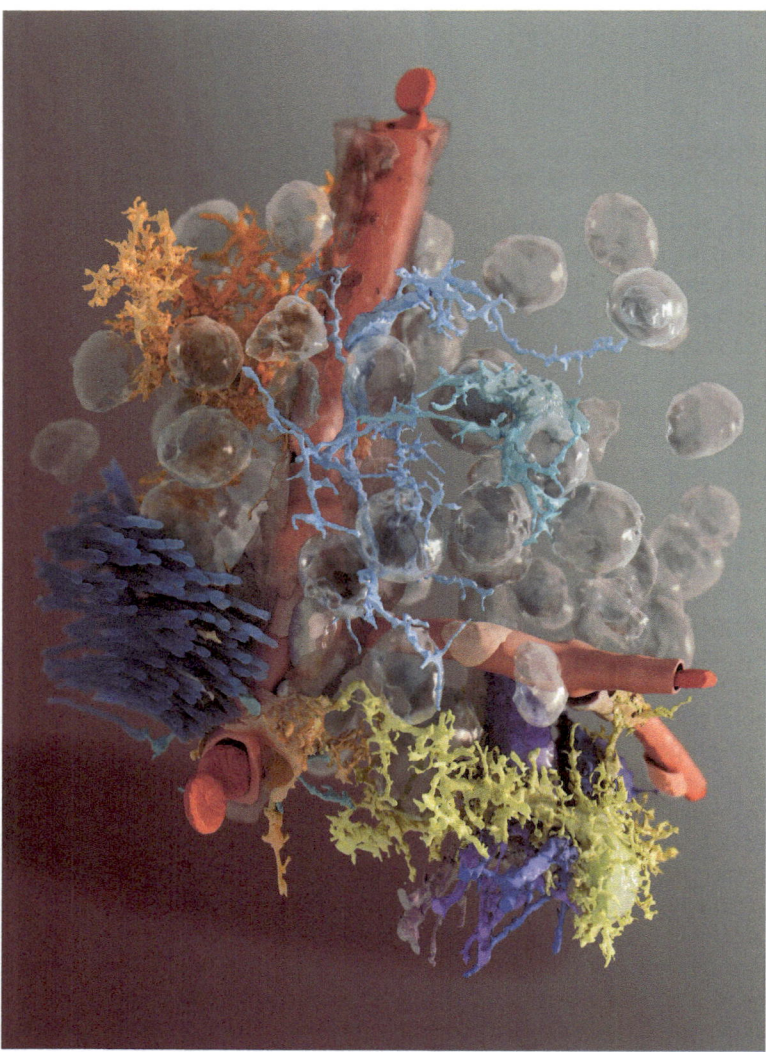

Fig. 4.1 The neuron in context. A digital 3D reconstruction of a micrograph stack from Layer VI of the somatosensory cortex of a P14 rat. Red: blood vessel; violet: neuron; light blue/green: microglia; dark blue: the fiber tract; orange (bottom): the pericyte; yellow/orange: astrocyte. Glass-like spheres: nuclei, reconstructed from the original image stack. (Adapted from Coggan et al., 2018, CC BY-NC 4.0 Deed)

While in the very beginnings of neuroscience, starting with the discovery of neurons in the brain, tremendous effort was applied to, technically and conceptually, isolate the neuron from its context, and to study its function separately, nowadays methods frequently used to register neuronal function, such as fMRI and PET, make use of these surrounding supportive structures and basic cell metabolism measures, for example, blood flow and glucose consumption, respectively. Accounting for this functional embeddedness of the neuron in its surrounding tissue, whenever possible, is crucial to further advance the field. Even when studying the core structural and functional features of the neuron, such as synaptic plasticity or neural cell metabolism, isolating the neuron leads to a reduced picture. Within the neuron, the cell-specific self-stabilizing feedback loops potentially underlying, for example, neural cell differentiation depend on the surrounding cells, primarily the supporting glia tissue. Functional specialization and its corresponding morphological and molecular changes are not simple properties of the neuron on its own. Functional specialization, stability, and plasticity are products of multiple interactions with the immediate and larger bodily context. In addition, any changes involve not only a single neuron but also groups of neurons, as well as their partly shared and partly distinct supporting glia cell structures. In particular, astrocytes function as integrative modulators across a large group of neurons, influencing synapse development, synaptic properties, excitability, and regionally specific gene transcription profiles (Allen & Lyons, 2018; Farhy-Tselnicker & Allen, 2018; Herrero-Navarro et al., 2021; Perez-Catalan et al., 2021). To put short a long argument, neurons are never alone. How do we account for this context in our understanding of the neuron and its function?

Cultural Contexts In Vitro

Important insights come from practical knowledge of how to successfully grow neurons in vitro. Research from the perspective of the History of Science and Science and Technology Studies repeatedly showed that experimental systems and research practices used in a specific field shape not only the theoretical models but also the empirical measurements collected with the applied methods (see Rheinberger, 1997, 2000, 2010; Soler et al., 2017; for neuroscience, see Choudhury & Slaby, 2012; for cell cultures, see Landecker, 2005, 2007). Case studies following this perspective often focus on the constraints and unspoken presumptions imposed on scientific objects and subjects by a specific research practice. They report unaccounted characteristics of the object or subject under study and unveil additional or counterknowledge not included in the scientific models. In the case of the neuron, we can use this focus on research practice for a first approximation of its functionally relevant context.

Neuronal cell cultures provide an extreme example of experimental reduction-ism, with the aim of reconstructing only the minimal necessary conditions for neu-rons to develop and survive outside a living organism. In an article reflecting on early research findings on the polarity of hippocampal neurons, Gary A. Banker (2018) recapitulates his long quest for the ideal cell culture medium. In the begin-ning, his neurons died after a few days, and only after years of trial-and-error exper-iments was his lab able to produce stable cell cultures that he could use in his experiments. He finally summarizes:

> [T]he fundamentals for successfully growing hippocampal cultures had been established: plate the cells at low density onto polylysine-treated glass coverslips, coculture them with astroglia growing on a separate substrate, and use N2 medium. With those key factors com-bined, we consistently obtained cultures that survived for a month or more, grew beautiful axons and dendrites, and underwent extensive synaptogenesis. (Banker, 2018, p. 1869)

In particular, the impact of growing astrocytes along with these neurons first went against his understanding of neuron development. "There was growing evi-dence at that time that a neuron's survival depended on molecular signals exchanged with its synaptic partners, so it seemed obvious to me that neurons in the explant were the source of trophic support" (Banker, 2018, p. 1868). It turned out, however, that this assumption was wrong. Experiments with astrocyte cultures[1] showed that these supported the neurons much better than the previously used explant cultures (Banker, 1980, 2018). Banker concludes, "[T]o my knowledge, the factors respon-sible for this effect have never been identified" (Banker, 2018, p. 1868). This report testifies to the nearly symbiotic relationship between neurons and astrocytes. Thus, in addition to the necessary growth medium, astrocytes provide a significant context for neural function.

Cultural Contexts In Vivo

While these in vitro cell cultures shed light on the minimal context necessary for a neuron to function properly, on the other side of the spectrum, *Cultural Neuroscience* investigates the role of the human-made sociocultural context for neuron activity (Kitayama & Tompson, 2010). This special branch of neuroscience studies, for example, the differences in brain structure and neural function between people with different cultural backgrounds. Prominently, Tang et al. reported distinct patterns of neural activation in Chinese native speakers, who were university students, and English native speakers, who were teaching English at the same university, during a simple mathematical calculation task (Tang et al., 2006). While the English native speakers showed a predominant activation of a semantic network that included the left perisylvian cortex, the Chinese native speakers showed predominant activation of a visuo-premotor association network. Based on their findings, Tang et al.

---

[1] Co-culturing the neurons with astrocytes was suggested and tested by Harry Kimelberg, a lab member at Banker's lab at the time (Banker, 2018).

theorized that the differences in the biological encoding of numbers and basic arithmetic calculations may be shaped by visual reading experiences during language acquisition and further cultural differences in math education. Specifically, they emphasize that the character-based language system of Mandarin trains a different pattern of semantic connections than the English language and that these different patterns are then used in the calculation process (Tang et al., 2006). Further findings of similar studies indicate differences in neural networks underlying different aspects of self-representation (Kitayama & Park, 2010, 2014; Sui et al., 2009), theory of mind processing (Kobayashi et al., 2006, 2007), and cognitive and emotional processing (for review, see Chiao, 2018; Kim & Sasaki, 2014).

While these approaches mostly understand culture as an independent exogenous factor that influences neural network development, some approaches emphasize an interactive understanding of the related culture–self relationship (Seligman et al., 2016). From this perspective, cultural traditions and practices, as well as cultural dimensions of body–mind concepts such as self, identity, and emotions, including biomedical knowledge inscribed in these concepts, shape our body–mind relationship in a way that potentially also reflects at a biophysiological and neural network level. When picked up in neuroscience research as cultural differences, the underlying neural correlates get integrated as part of the bio-cultural constitutions of sociocultural groups—a process which Seligman and others termed "bio-looping" (Seligman et al., 2016). For example, as Suparna Choudhury showed, studies investigating the adolescent brain built on the Western notion of adolescence for the identification of adolescent-specific neural correlates and, by doing so, further manifested the notion of adolescence as a biologically distinct phase in human ontogenetic development (Choudhury, 2010). Adolescence, however, when studied from a historical and cultural-comparative anthropological perspective, did not occur before the twentieth century (Choudhury, 2010). Accordingly, Seligman, Choudhury, Kirmayer, and others proposed adding the analysis of the bio-looping effects of cultural knowledge and resources to Cultural Neuroscience research, including the study of scientific concepts and how they influence the active construction of self-perception patterns and identities, emotion regulation, and health and body-related behavior (Choudhury, 2010; Kirmayer, 2011; Kirmayer & Gómez-Carrillo, 2019; Seligman, 2017; Seligman et al., 2016).

These examples of Cultural Neuroscience findings and concepts show that differences due to the cultural context of psychobiological development can be visualized at the neural activity level, that the visualization of neural substrates creates new sociocultural subgroups, such as the neurally distinct group of adolescents, and that these provide new forms of sociocultural conditions that constitute psychosocial developmental contexts that then leave traces at the neural activity level. This bio-cultural looping effect demonstrates not only the tremendous plasticity and sociocultural entanglement of our brain. It also provides further evidence that the underlying bio-cultural coordination processes are more than just sight effects. All points to the fact that these bio-cultural interactions are constitutive for the development of proper mental functions, and we can rightly assume that they have correlates at the molecular level. In addition, these studies show that specific psychological functions do not develop alone, but are influenced by and co-develop with other functions and capabilities, such as language processing, reading, mathematical calculations, self-perception, and self- and group identification. Complex psychological functions and their related network connections provide a specific context for other psychological functions and associated neural activity, even at lower levels, such as in sensory processing and all the way down to the single neuron. Accordingly, neural networks provide an important context for single-neuron activity.

The Neural Network Context

In the old neuron theory, the firing rates of single neurons were interpreted as a unique language in which all sensory input and motor activity are encoded in the brain. In this view, direct dendritic and synaptic connections between neurons were, if not the only, the most important context of a neuron's activity, enhancing or depressing action potentials in the postsynaptic target neuron. The neuron was seen as a reliable, but also static, if not stoic, producer of this action potential once the threshold for its activation was reached. The underlying model describes fairly good how neurons produce action potentials in vitro and isolated from most of their complex neural and other tissue connections. However, in vivo, things are fuzzier.

The encoding of information, according to the "old view," was determined by frequencies of firing rates, synaptic connectivity, and specialization. This view has fundamentally changed in recent years. Gerstner et al. characterized "new," emerging perspective as "population-based" and outlined the consequences for our understanding of the neuron as follows:

> While early models of cortical processing suggested that individual neurons are sufficient to represent detailed descriptions of relevant features of the environment, it is now generally accepted that unambiguous representations are based on population codes. (Gerstner et al., 1997, p. 12740)

Thus, "the firing rate of individual neurons in general cannot provide an unambiguous description, even for simple stimuli" (Gerstner et al., 1997, p. 12740).

Instead, as the authors emphasize, neuroscientists need to acknowledge a strong "history dependence" of action potentials and that the effect of presynaptic signaling on the postsynaptic neuron varies from action potential to action potential, even between direct connections (Gerstner et al., 1997, p. 12740). Gerstner et al. (1997) broadly differentiate three types of frequency dependent synaptic signals, which need to be accounted for: While depression and facilitation of signal transmission are assumed to depend on (1) timing and (2) amplitude, a third type of signal transmission seems to rely on (3) the number of presynaptic action potentials—as if the synapses "count" the incoming action potentials. Consequently, the signal transmission varies significantly depending on the state and characteristics of the postsynaptic neuron. Gerstner et al. concluded that "each synapse selects a unique mélange of features of the presynaptic AP [action potential] train and transmits only a specific subset of the information contained in the entire train. Different aspects of the same train are read out by different target cells" (1997, p. 12740). They also emphasized that the interaction and co-activation of neurons, in the way that action potentials are "phase locked," depends on fine-tuning mechanisms during early developmental periods as well as the immediate prior activation history (Gerstner et al., 1997, p. 12741). From a developmental perspective, this selectivity allows for some fuzziness of information processing and encoding at the single-neuron level, with the single neuron focusing on some aspects of the incoming signals and treating others as background noise. With several neurons reacting to different aspects of the same signal, the signal is represented at a greater depth at the population level, with single neurons overlapping in their detection bandwidth. The involvement of neuronal populations in different encoding processes is an important mechanism for ensuring the accuracy and stability of information processing in the brain. Consequently, a single-neuron activity must always be interpreted within this population context.

From a methodological standpoint, Gerhard et al. (2011) later called this change in perspective a "paradigm shift in the analysis from single-spike statistics toward the analysis of neural population activity and interactions between neurons." This shift was made possible by new technical and computational advances in electrophysiology and systems neuroscience, which allowed the recording of many neurons in parallel (e.g., Georgopoulos et al., 1986; Jarosiewicz et al., 2008; Nicolelis et al., 1997; Santhanam et al., 2006). The population-based view accounts for the indeterminacy and equifinality of neuronal activation seen, for example, in the neural basis of the song-singing behavior of zebra finches discussed in the previous chapters, which simultaneously provides stability of function and flexibility at the single-neuron level at the same time. It also points to the fact that the context of the neuron at the level of neuronal networks has not only a spatial component but also a time component (Jurjuţ et al., 2011). The spatial component, which accounts for all physiological connections via synapses and dendrites, long-distance connections with interneurons, and all types of cell–cell communication within the local cellular milieu, still corresponded with a single neuron–based view. However, the time component is incompatible with this view's inherent reductionism. The short- and long-term developmental history of the neuron, comprising cell differentiation history, connectivity history, and more recent activation history, points to changing activation patterns at the single-neuron level. Thus, considering the time component unveils the population-based nature of functional stability and specialization. In this, the "cell memory," represented, for example, by relatively stable transcriptional patterns which are likely modulated and maintained at several levels including the epigenetic activity level, plays an important role. In addition, the time component also relates the cell memory and the immediate network context with other contexts, such as the direct cellular surroundings, the body and its most basic as well as more complex bodily functions, and the sociocultural environment relevant for psychobiological development. All these factors provide and constitute the broader developmental context of the single neuron.

Within the field of computational neuroscience, we observe attempts to integrate the time component of the single-neuron activity in addition to the spatial component within statistical models (see, e.g., Gerhard et al., 2011; Kasabov, 2018). One approach is to capture global network activities (Brown et al., 2004; Iyengar, 2003; Paninski et al., 2010; Pillow et al., 2008; Truccolo et al., 2005). Other approaches model the input at the single-neuron level in dependence of local measurable modulators. Gerhard et al. (2011), for example, proposed to account for the neuron's "self-history," the input from other neurons, and its modulation by external stimuli. However, approaches that combine the spatial and temporal components of neural activity are rare. Most network attempts to model population activity based on single-neuron activity still investigate either the spatial or temporal component of information encoding in the brain, while the other is more or less neglected (Jurjuţ et al., 2011). The main reason for this is the computational complexity of the attempt. Combining spatial and temporal information at the level of a single neuron to calculate the activity at the neuron population level is computationally intensive. As Beniaguev et al. (2021) demonstrated in a simulation study, a deep neural network with five to eight layers is necessary to model the activity of a single neuron, a layer 5 cortical pyramidal cell, based on the known biophysical properties of its dendritic connections and synaptic function. This clearly shows how much complexity but also fuzziness results from the multiple input–output relationships in which a single neuron is involved, even when only focusing on the direct connections with other neurons.

As discussed in the previous chapter, Gregor Schöner et al. argued that because of its fuzziness, local instability, and recursive character, the activity of a single neuron is not informative enough to relate neural activity to behavior and psychological functions, and that we should ignore it and replace it with measures of population activity. In this way, they avoid the tedious task of having to model the complexity at the single-neuron level and simply treat the single-neuron activity as a black box. An alternative approach to reduce complexity is to focus on one property of neuron activity that captures the single neuron as well as the population level in its spatial and temporal dimensions of signaling. Some rhythmic patterns across neurons, described as "synchronicity," are used in this way.

"In Sync" with the Network

In the 1980s, Christoph von der Malsburg introduced the notion of synchronicity to solve the coding problem in sensory processing. Based on his empirical studies of the visual system, he proposed that the integration of different aspects of a visual stimulus, such as color, location, shape, and movement, uses temporal information in the activity patterns of the participating neurons (von der Malsburg, 1994). He argued that the functional specialization of single neurons alone cannot explain the multiplicity of representations based on the combined features, as the number of possible variations would by far exceed the number of neurons. Synchronization allows a single neuron to participate in different lower-order temporally coordinated neural assemblies or networks, making it possible for a large number of states to be encoded by a small number of neurons. Von der Malsburg's work on synchronicity was later continued, among others, by Wolf Singer, who investigated how time-related aspects of sensory stimuli were processed in the visual cortex of cats (Gray & Singer, 1989). Singer defined neural synchronicity as the coordinated firing rate of neurons (Singer, 1999). Accordingly, perfect synchronicity represents the case in which all neurons of that population fire at exactly the same time point and no synchronicity means that they fire at different time points than other neurons in that population, while intermediate grades of synchronicity mean that either a fraction or fractions of that population are firing together or in a time-related manner.

The strength of this approach is that the focus on timing in synchronicity allows for the collection of measurements at different activity levels that are also translatable across levels. At the individual neuron level, synchronicity is measured using cross-correlograms of spike trains and their correlation with the local field potential oscillations (Friston, 1997). At the cortical column level, synchronicity represents the coordination of the postsynaptic activity of networks of pyramidal cells, which gives rise to an electrophysiological signal measurable via EEG (Barr et al., 2009). The power of these signals represents the synchronized activity of the underlying networks. However, within this coupled synchronized activity at the population level, there can still be functionally relevant asynchronicity at the level of the single neuron. Thus, the context of a single neuron is multidimensional in terms of synchronicity across different levels. Accordingly, at the single neuron level, Singer points out that temporal coordination represented by synchronicity "is not a trivial reflection of anatomical connectivity such as shared input through bifurcating axons, but instead results from context-dependent, dynamic interactions within the cortical network" (Singer, 1999, p. 49). Most importantly, this line of research has shown that at least some of the coordination processes to achieve synchronicity are not superimposed by higher network levels but are realized at the level of small neural networks in groups of neurons encoding low-level features, for example, visual stimuli.

Based on these findings, Singer and colleagues proposed the so-called temporal-binding hypothesis (Engel et al., 1999; Singer, 1999, 2021) according to which the information which aspects of the perceived environment need to be integrated is encoded by this low-level temporal coordination. This hypothesis questioned the assumed multilevel architecture of the CNS and replaced the notion of a clear-cut separation of sensory areas, a processing control center, and executing areas with the notion of many processing hubs. No central unit or center integrates the incoming information. Instead, integration is assumed to take place at all levels of the system based on the so-called small world networks: "Small world networks, of which the brain seems to be an example [...], combine these two processing modes: local modularity and long-range connectivity. Thus, depending on dynamic shifts in coupling, the same anatomical network can support both local and global—or subconscious and conscious—operations" (Uhlhaas et al., 2009, p. 10).

Evidence of the importance and ubiquity of synchronicity in the brain is based on a wide range of findings. Inhibitory interneurons have been shown to be crucial for the rhythmic pacing of neuronal activity across longer distances (Cardin et al., 2009; Sohal et al., 2009; for a review, see Bartos et al., 2007). In addition to the local field potentials of neural populations, brain rhythms within the background neuronal activity at specific frequencies are also generated as coordinated outputs of larger networks (Barr et al., 2017), and their therapeutic modulation can improve impaired cognitive processing (Grover et al., 2021). Accordingly, a potential functional role of this low-level synchronicity observed in sensory processing for consciousness, neural diseases, and psychiatric disorders has been proposed (Uhlhaas et al., 2009), although there is also disagreement as to whether the temporal-binding hypothesis holds even for basal sensory processing. For example, Shadlen and Movshon (1999, p. 69) argue that "a fairly high level in the visual cortical hierarchy" needs to participate in binding the different aspects of a visual stimulus together, and that synchronicity phenomena at lower levels are mainly imposed by these higher binding processes.

At the single-neuron level, the local and large-scale synchronicity–asynchronicity patterns are part of a vast stream of information that connects the single neuron to smaller and larger levels of neural interaction. This stream of information reaching and leaving the single neuron entails multiple layers of signal and noise and needs to be decoded to be functionally relevant (Bialek et al., 1991). In this sense, one could say, "Neurons are noisy information-processing elements" (Malik & Ajemian, 2017, p. 23). The spike activity of a single neuron depends largely on the local and larger network contexts. It also depends on the developmental and activation history, which is embedded within these local and larger contexts. Together, these provide the conditions under which a single neuron participates in neural activity. However, from the perspective of the single neuron, the different inputs need to be coordinated and balanced according to its cellular state and cell memory in a way that is energy efficient, fast to decode, relatively stable, and still open for change. Here, epigenetic mechanisms come into play.

The Epigenome as Signal-Integration Platform

It has been discussed for quite some time, that epigenetic mechanisms are involved in when neurons participate in neural networks, that they play a functional role in memory and learning, and that their dysfunction may impact mental health (Gräff & Mansuy, 2008; Isles, 2015; Isles & Wilkinson, 2008; Levenson & Sweatt, 2005; Marshall & Bredy, 2016; Mews et al., 2021; Sweatt, 2013). First hints came from the finding that the CREB-binding protein (CBP), which is required for long-term memory formation and enhances the binding of CREB to the genome (Dash et al., 1990; Lonze & Ginty, 2002), functions as histone acetylase (Guan et al., 2002). Studies on DNA methylation changes in memory-related brain areas, such as the hippocampus, with fear conditioning learning paradigms further emphasized the direct participation of epigenetic mechanisms in neural activity regulation (Day & Sweatt, 2011; Levenson et al., 2006; Miller et al., 2008; Miller & Sweatt, 2007). Some researchers even hypothesized that epigenetic mechanisms participate in the formation of *engrams*, which are neuronal ensembles that form long-lasting connections related to individual memories, by modulating the initiation and maintenance of the underlying neural connections (Kyrke-Smith & Williams, 2018; Ripoli, 2017).

In a review on epigenetic mechanisms involved in the regulation of synaptic functions, Campbell and Wood state,

> researchers have turned to the epigenome as a signal-integration platform through which neurons might integrate new information at the molecular level in the service of stable changes in cell function. (2019, p. 133)

However, it is not addressed if this integration process is a single-cell process or a coordinated group process of neighboring neurons and glia cells. In the tradition of the classical neuron theory, most studies aim to reconstruct processes at the single-neuron level, although whole tissue samples are often used to estimate neuron-specific epigenetic modifications (see Chap. 5 for a detailed discussion of this issue).

Campbell and Wood (2019) reviewed the initial findings on synapse–epigenome, dendrite–epigenome, and epigenome–synapse signaling. These indicate that epigenetic mechanisms are involved in the activity-induced remodeling of synapses and dendrites by supporting the binding of transcription factors or regulating the transcription of supporting enzymes. The most well-studied example is the translocation of the CREB-regulated transcriptional co-activator 1 (CRTC1) from the dendrites to the nucleus, where it assists in the regulation of a set of CREB target genes. Most importantly, this regulation may, in part, be sensitive to the duration and intensity of activation. For example, in a study on contextual fear memory in mice, Uchida et al. (2017) showed that the expression of brain-specific fibroblast growth factor 1 B (FgF1B) in the hippocampus depends on the presence of CRCT1 in the nucleus. Thirty minutes to 1 hour after weak or strong associative training, with the mice receiving one or three electroshocks, the HDAC3-nuclear receptor co-repressor complex was removed from the *Fgf1b* promoter, and phosphorylated CREB and CBP were recruited to induce transient expression. Two hours after strong training but not after weak training, Uchida et al. observed a CRTC1-mediated exchange of CBP for the histone acetyltransferase KAT5 at the promoter of *Fgf1b*, which induces persistent changes in *Fgf1b* expression and enhances long-term memory. In a different series of studies, Van Leuween et al. (2014) reported that the synaptic protein afadin shuttles to the nucleus to promote the phosphorylation of H3 at serine 10 (H3S10), which leads to a more permissive state of densely packed chromatin, enabling transcription. Although probably a more general mechanism—specific gene loci affected by this mechanism are still unknown—the findings indicate that H3S10 phosphorylation is required for dendritic spine remodeling.

For other proteins, the mechanisms that transport information from the synapses are not well understood. However, there are indications that more synapse-related proteins are regulated by epigenetic mechanisms. One example is the involvement of the nuclear histone lysine methyltransferase G9a in the degradation cycle of AMPA receptors. After exposure to stress, glutamatergic AMPA receptors are degraded by ubiquitination. Wei et al. (2016) reported that in rats repeated stress led to glucocorticoid-receptor-dependent increases in the expression of HDAC2, which then increased the occupancy of HDAC2 at the promoter of *G9a* and reduced G9a expression. Normally, G9a methylates histones at the promoter of *Nedd4*, a gene encoding an E3 ubiquitin ligase, thereby reducing its expression and potentially increasing the likelihood of AMPA receptors being ubiquitinated. Thus, G9a participates in fine-tuning AMPA receptor degradation under stress conditions. This shows that, in addition to neural activity, stress hormone signaling affects synaptic regulation via an epigenetic regulator (HDAC2), which persistently alters synaptic receptor states. It also shows that, at least in the case of histone modifications involved in synaptic mechanisms, the epigenetic state of the involved gene loci can be highly dynamic and needs to be actively maintained within the respective transcription cycles.

Most of the epigenetic modifications involved in synaptic regulation are histone modifications that modify the accessibility of chromatin as well as the binding of transcription factors. Some of them act locally, targeting only a specific type of binding site, as in the case of CBP. Others show potential for an overly broad function, affecting the epigenome as a whole or at least its more accessible parts, such as the synaptic or dendritic regulation of epigenetic functional enzymes involved in DNA methylation and histone modifications (HATs, HDACs, DNMTs, etc.) as, for example, those found for GluN2A-containing N-methyl-D-aspartate receptors (NMDAR) (Bayraktar et al., 2020; Bayraktar & Kreutz, 2018). Furthermore, some epigenetic modifications related to synaptic activity appear to depend on the intensity of neural activation. When observed in the context of learning, this indicates that epigenetic regulation takes part in immediate and long-term adaptations of the single neuron to the signaling context. Overall, these first known modifications indicate that there are several synaptic epigenetic mechanisms participating in the regulation of synaptic activity and plasticity, both short- and long-term, and likely more will be discovered. Thus, the activity of the surrounding neural networks induces short- and long-term epigenetic modifications within the single neuron, which then changes the way the single neuron participates in signaling processes within these networks (for an overview of the potential epigenetic signaling pathways, see Fig. 4.2). In addition to activity-dependent epigenetic modifications induced via dendritic or synaptic signaling, there is also first evidence for epigenetic modifications induced via hormonal signaling (Zhang & Ho, 2011) or direct cell–cell signaling between neurons and glia cells (Wang et al., 2020). Both neural activity-dependent and neural activity-independent (e.g., hormonal, homeostasis-related, or glia signaling) induced epigenetic modifications likely interact with each other (for discussion, see Lux, 2018), providing a battery of complex regulatory mechanisms of neural maintenance and function.

Taking up the notion of the epigenome as a "signal-integration platform" (Campbell & Wood, 2019), the question arises as to how these different pathways

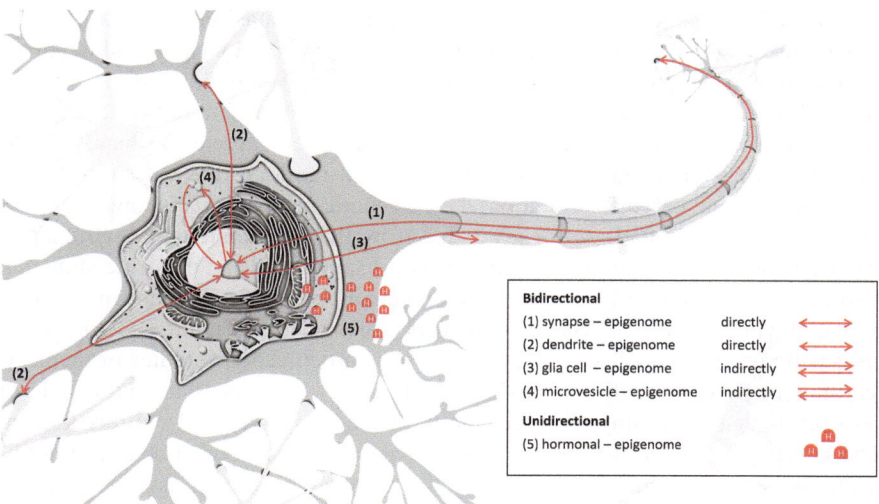

Fig. 4.2 Signaling pathways connecting external input to the epigenome within neurons. Bidirectional signaling to and from the epigenome occurs via (1) axon synapses, (2) dendrites, (3) glia cells, and (4) microvesicles. Signaling from and to axon synapses and dendrites is directly bidirectional, with the origin and target of the signal staying the same in both directions. Signaling from glia cells and microvesicles is indirectly bidirectional, with the origin and target of the signal being of the same type but not exactly the same cell or vesicle. Unidirectional signaling to the epigenome occurs via (5) hormonal signaling

shape the individual epigenome of a neuron. Several layers of epigenome-modifying feedback loops are probably initiated and maintained by these signaling pathways, serving different functional purposes in a neuron, from cell differentiation and homeostasis to cell protection and functional specification. However, with the currently available methods, it is not trivial to disentangle these layers. Despite these first steps, there is still a long way to go. Biostatisticians have already successfully predicted cell types, including neurons and glia cells (Rivera & Ren, 2013; Scott et al., 2020; Wytock & Motter, 2020), and in some cases of cancer cells, even individual cell lines and cell states (Capper et al., 2018; Danielsson et al., 2015; Koelsche et al., 2021) from epigenetic patterns. Human epigenome projects published common epigenomic identifiers for several cell types (Beck, 2014; Kundaje et al., 2015), which allow, for example, the prediction of cell-type compositions in tissue samples (Zhu et al., 2022). However, the neuron-specific epigenetic patterns extracted from these epigenomic data streams represent only one layer of the epigenome of a single neuron. Individual developmental pathways and connectivity patterns of neurons or neuron–glia cell ensembles contribute to the differentiation and individualization of the single-neuron epigenome. At the current state of knowledge, we need to assume that epigenetic mechanisms are a crucial part of a neuron's functional individualization process.

Some of these signaling-induced epigenetic modifications serve as synaptic epigenetic mechanisms. Primarily, these are those induced by synaptic and dendritic inputs, but neuron–glia cell communication, which plays a modulating role in synaptic activity, also involves epigenetic mechanisms. Furthermore, some of these epigenetic modifications function as developmental epigenetic mechanisms that impact the developmental pathways of neurons in the long run. During the early phase of neuronal cell differentiation, the epigenetic mechanisms involved are less functionally specific and are locally induced by the immediate cell environment. After a neuron's integration into neural networks, synaptic and dendritic signaling also potentially induces long-term developmental epigenetic mechanisms. In particular, the point at which synaptic epigenetic mechanisms turn into developmental epigenetic mechanisms is of great interest for neuroepigenetic studies of memory, learning, or psychiatric diseases (Mews et al., 2021). A major issue in this field of research is that the timing of epigenetic modifications is still not entirely known. The same holds for sensitive periods, during which we see a higher degree of modifications being stabilized over time in the affected neuron or neuron–glia cell ensembles. First studies indicate that early developmental windows represent sensitive periods for functionally relevant modifications with long-term effects (Li et al., 2022; Martins et al., 2021), including cell-type-specific effects (Rahman & McGowan, 2022), but further investigation of such patterns over the lifespan is needed. Knowing the timely coordination of epigenetic modifications is an important prerequisite for effectively tracing them in an experimental context or even in a clinical study.

Thus, a neuron's epigenome provides us with dense layers of information on a neuron's individualized developmental history as well as its current activity state. The larger task is to register and interpret this multilayered information. However, even if we could decipher all its details from the perspective of function, this individualized single-neuron epigenome only reports the potential for participation of a single neuron within its functional network or the incoming stream of signaling patterns. It neither provides a stored blueprint of the function previously executed or executed in the future nor an exact functional history. As it becomes increasingly clear, function actively acquires its necessary neuronal resources from a population of neurons. Among these, a single neuron is only one of the thousands of potential neurons participating in this moment to bring about the function. This bidirectional notion of function—that function initiates its own neural foundation, which then carries it—is not easy to grasp conceptually. In particular, it means that determining which neurons participate in an upcoming function will not be possible before the function itself is realized. In parallel, epigenetic modifications do not represent the inscribed instructions of function. The epigenome only represents a rough marker for the single neuron's developmental and activation history in the form of a dynamic epigenetic cell state that integrates different signaling pathways. In addition, this epigenetic cell state only reflects signals that reach a threshold for inducing epigenetic modifications. At best, we will be able to use this marker as a proxy for the bandwidth of the functional specialization and plasticity range of a neuron, group of neurons, or neuron–glia ensembles at a specific time point.

The Neuron as Product and Producer of Its Context

While the epigenetic cell state potentially contains information about contextual influences that shape a neuron's functionality, it does not provide a measure of function at a higher functional level of neural activity, at least not when measured at the single-neuron level, which is considered the future gold standard in the field (Rahman & McGowan, 2022). The reason for this lies in the relationship between the single neuron and higher functional levels. Our growing knowledge of the plasticity of neural function and the neuron itself indicates that an exact neuron function relationship is a biological exception in the human brain. This insight is the unintended result of improving measurements and technological innovations, including those registering the single-cell epigenome, which allow us, with every step, an even deeper look into the workings of the brain. The better we can paint the picture of the single neuron, the more we observe its individuality and canalized flexibility. Moreover, this canalized flexibility produces a blurred picture that becomes fuzzier as we try to pinpoint a neuron's functional involvement in a specific behavioral action or mental state. The old neuron theory, according to which all neural activity has its *starting point* within single neurons, no longer holds. We need to reconsider our understanding of the neuron in light of this indeterminacy, and develop new concepts and ultimately a new neuron theory. This new neuron theory must first account for the tremendous flexibility of neuronal activation. Second, it must account for the bidirectionality of neural connections, due to which changes in a neuron's structure and function potentially induce changes in its neural networks or other signaling pathways. The neuron's activity is induced by its context and induces neural activity within this context. Third, the neuron is strongly affected by its developmental and activation history—its lifespan context—which results in highly individualized activation patterns, even when comparing neighboring neurons. Flexibility, bidirectionality, and developmental history depend on each other, canalizing the potential structure and function of the neuron.

These characteristics of the neuron are of no surprise when we recapitulate the functional role of the nervous system from an evolutionary perspective: to reflect and interconnect the external environment and the internal state of an organism (Leontiev, 1981). Within a constantly changing environment, only excitable neurons with canalized synaptic flexibility can successfully perform this task. The structural and functional plasticity range of a neuron is wider than that of other cells, which is clearly one of its key characteristics. However, from a biological perspective, this flexibility requires stabilization to make sense energetically. What we observe and measure as a function of a single neuron is the result of these stabilization processes at a certain time point. Network phenomena, such as synchronicity and their role in integrating neurons in local and distant neural networks, play an important role in these stabilization processes. Also, as discussed, glia–neuron signaling participates in several aspects to these processes. Finally, neuroepigenetic mechanisms have biophysical properties that make them strong contributors to these stabilization processes. We already know that they canalize and enable development at the individual cell level, maintain a specific activation history and state in correspondence with structural-genomic epigenetic mechanisms, and coordinate basic transcription loops necessary for the neuron's metabolism and function. In particular, this coordinative role, balancing metabolic and functional affordances, requires further investigation.

Thus, by inducing, canalizing, and stabilizing its activity, the context produces the neuron and its function. However, the single neuron is not a passive receiver of signaling information but actively responds depending on its developmental and activation history. This developmental and activation history is also the result of the context, yet a past context, which is maintained as information within the single neuron and its signaling context. Moreover, this past context is not limited to the immediate cell environment within the brain, but also includes the permanent developing and changing brain as an organ system, the life history of the organism, and, in the case of our human mind and brain, its sociocultural embeddedness, as we have seen in Cultural Neuroscience studies.

From the perspective of function, stabilization via network integration is the most obvious. As long as a group of neurons is integrated into a signaling network for a specific function, the activity patterns are strongly influenced by this network. However, the functional specificity can only be in part the result of the network, and this part is highly temporary. Even at the smallest local network level, there is no complete correspondence between the single-neuron activity and activity at the neural population level.

We can also consider this as a form of an inbuilt backup system. The organism cannot afford to store vital functions in only one cell. There is always a high risk that this single cell will die, and with it, its specialized function. However, this also means that memory, for example, cannot be purely synaptic, although it has a synaptic component. It would be very inefficient if a single neuron loses all its functional specificity when it is on pause and not integrated in the immediate signaling pathway. Therefore, additional backup mechanisms at the single-neuron level, as well as the higher network levels, need to be in place. At the single-neuron level, neuroepigenetic mechanisms provide the means for such an energy-efficient individual neuronal cell memory. Once a formerly abandoned neuron is newly addressed by a signaling network, the epigenetic cell state enables a faster and more efficient second integration. However, this means that we need to understand neuroepigenetic mechanisms, as well as those in the supporting cell structures, as mechanisms that stabilize function and enable plasticity by constraining flexibility, as outlined in the previous chapter. This neuroepigenetic memory of the individual neuron's trajectory also results from, and thus reflects, the neuron's inter- and co-action with its context.

It is important to note that neuroepigenetic memory is not an engram of specific memory content. It has no "meaning tags." Neuroepigenetics provides us with additional data to register neural functions in a content-independent manner. The content of thought processes is accessible only through introspection. However, at the level of neuroepigenetic mechanisms, only that something is remembered—and not what—can be traced as well as the molecular requirements of such memory. This is not further illuminating when everything works, but is of high interest when memory processes fail or are just not working properly, such as in dementia and Alzheimer's disease. Studying the role of neuroepigenetic mechanisms in psychobiological development would then mean to ask whether the same or different neuroepigenetic mechanisms contribute to memory processes and their function, as well as dysfunction related to learning the first language, learning to ride a bike, remembering traumatic events, and not remembering where I put the keys in old age (early stages of dementia). However, the epigenetic cell state is not a perfect account of a neuron's past and presence, but only a fuzzy trace that roughly shapes its current activity status and signaling sensitivity.

This indeterminacy is difficult to endure at the single-neuron level. At higher levels, it seems much easier for us to tolerate uncertainty in the material basis of neural function. The fact that the correspondence between behavior and neuronal basis is not perfect seems much more obvious at the sensory and motor activity level compared to lower neural activity levels. In addition, the emerging connectivity paradigm and the success of modeling approaches relying on systems neuroscience already established a change in perspective for these higher levels. The indeterminacy is also captured in the methodology for downstream molecular and biophysical processes at the subcellular level, for which the use of threshold models and system modeling of feedback loops is also fully established. At the single-neuron level, the old neuron theory stands in the way of a notion of the neuron that is open to the context dependence of neural function, as reflected in the measurements and experimental findings. Cajal's picture of the stained neuron is still too powerful.

However, the neuron is never alone. It is not only the producer, but also the product of neural function. Moreover, its functionally relevant context extends from the immediate cell environment to our extensive human culture, and this context produces the neuron in a double sense: as a functional cell with a developmental history within an organism–environment system, and as a target of our scientific efforts to trace its function. In this sense, the neuron is the product of our bio-sociocultural system. To overcome our simplified view of the single neuron, we must reintegrate our neuron theory into a broader conceptualization of psychobiological development. I discuss this further in the last chapter. First, I will characterize in more detail what we can trace at the level of epigenetic activity.

References

Allen, N. J., & Lyons, D. A. (2018). Glia as architects of central nervous system formation and function. *Science (New York, N.Y.), 362*(6411), 181. https://doi.org/10.1126/science.aat0473

Arber, S. (2017). Organization and function of neuronal circuits controlling movement. *EMBO Molecular Medicine, 9*(3), 281–284. https://doi.org/10.15252/emmm.201607226

Bacci, A., Verderio, C., Pravettoni, E., & Matteoli, M. (1999). The role of glial cells in synaptic function. *Philosophical Transactions of the Royal Society B: Biological Sciences, 354*(1381), 403–409.

Banker, G. A. (1980). Trophic interactions between astroglial cells and hippocampal neurons in culture. *Science, 209*(4458), 809–810. https://doi.org/10.1126/science.7403847

Banker, G. A. (2018). The development of neuronal polarity: A retrospective view. *Journal of Neuroscience, 38*(8), 1867–1873. https://doi.org/10.1523/JNEUROSCI.1372-16.2018

Barr, M. S., Farzan, F., Rusjan, P. M., Chen, R., Fitzgerald, P. B., & Daskalakis, Z. J. (2009). Potentiation of gamma oscillatory activity through repetitive transcranial magnetic stimulation of the dorsolateral prefrontal cortex. *Neuropsychopharmacology, 34*(11), Article 11. https://doi.org/10.1038/npp.2009.79

Barr, M. S., Rajji, T. K., Zomorrodi, R., Radhu, N., George, T. P., Blumberger, D. M., & Daskalakis, Z. J. (2017). Impaired theta-gamma coupling during working memory performance in schizophrenia. *Schizophrenia Research, 189*, 104–110. https://doi.org/10.1016/j.schres.2017.01.044

Bartos, M., Vida, I., & Jonas, P. (2007). Synaptic mechanisms of synchronized gamma oscillations in inhibitory interneuron networks. *Nature Reviews. Neuroscience, 8*(1), 45–56. https://doi.org/10.1038/nrn2044

Bayraktar, G., & Kreutz, M. R. (2018). Neuronal DNA methyltransferases: Epigenetic mediators between synaptic activity and gene expression? *The Neuroscientist, 24*(2), 171–185. https://doi.org/10.1177/1073858417707457

Bayraktar, G., Yuanxiang, P., Confettura, A. D., Gomes, G. M., Raza, S. A., Stork, O., Tajima, S., Suetake, I., Karpova, A., Yildirim, F., & Kreutz, M. R. (2020). Synaptic control of DNA methylation involves activity-dependent degradation of DNMT3A1 in the nucleus. *Neuropsychopharmacology, 45*(12), Article 12. https://doi.org/10.1038/s41386-020-0780-2

Beck, S. (2014). The human epigenome project: Past, present, and future. In *Reference module in biomedical sciences*. Elsevier. https://doi.org/10.1016/B978-0-12-801238-3.00096-9

Beniaguev, D., Segev, I., & London, M. (2021). Single cortical neurons as deep artificial neural networks. *Neuron, 109*(17), 2727–2739.e3. https://doi.org/10.1016/j.neuron.2021.07.002

Bialek, W., Rieke, F., de Ruyter van Stevenick, R. R., & Warland, D. (1991). Reading a neural code. *Science (New York, N.Y.), 252*(5014), 1854–1857. https://doi.org/10.1126/science.2063199

Brown, E. N., Kass, R. E., & Mitra, P. P. (2004). Multiple neural spike train data analysis: State-of-the-art and future challenges. *Nature Neuroscience, 7*(5), Article 5. https://doi.org/10.1038/nn1228

Campbell, R. R., & Wood, M. A. (2019). How the epigenome integrates information and reshapes the synapse. *Nature Reviews. Neuroscience, 20*(3), 133–147. https://doi.org/10.1038/s41583-019-0121-9

Capelli, P., Pivetta, C., Soledad Esposito, M., & Arber, S. (2017). Locomotor speed control circuits in the caudal brainstem. *Nature, 551*(7680), Article 7680. https://doi.org/10.1038/nature24064

Capper, D., Jones, D. T. W., Sill, M., Hovestadt, V., Schrimpf, D., Sturm, D., Koelsche, C., Sahm, F., Chavez, L., Reuss, D. E., Kratz, A., Wefers, A. K., Huang, K., Pajtler, K. W., Schweizer, L., Stichel, D., Olar, A., Engel, N. W., Lindenberg, K., et al. (2018). DNA methylation-based classification of central nervous system tumours. *Nature, 555*(7697), Article 7697. https://doi.org/10.1038/nature26000

Cardin, J. A., Carlén, M., Meletis, K., Knoblich, U., Zhang, F., Deisseroth, K., Tsai, L.-H., & Moore, C. I. (2009). Driving fast-spiking cells induces gamma rhythm and controls sensory responses. *Nature, 459*(7247), Article 7247. https://doi.org/10.1038/nature08002

Chiao, J. Y. (2018). Developmental aspects in cultural neuroscience. *Developmental Review: DR, 50*(A), 77–89. https://doi.org/10.1016/j.dr.2018.06.005

Choudhury, S. (2010). Culturing the adolescent brain: What can neuroscience learn from anthropology? *Social Cognitive and Affective Neuroscience, 5*(2–3), 159–167. https://doi.org/10.1093/scan/nsp030

Choudhury, S., & Slaby, J. (2012). *Critical neuroscience: A handbook of the social and cultural contexts of neuroscience*. Wiley-Blackwell.

Coggan, J. S., Calì, C., Keller, D., Agus, M., Boges, D., Abdellah, M., Kare, K., Lehväslaiho, H., Eilemann, S., Jolivet, R. B., Hadwiger, M., Markram, H., Schürmann, F., & Magistretti, P. J. (2018). A process for digitizing and simulating biologically realistic Oligocellular networks demonstrated for the neuro-Glio-vascular ensemble. *Frontiers in Neuroscience, 12*. https://www.frontiersin.org/articles/10.3389/fnins.2018.00664

Danielsson, A., Nemes, S., Tisell, M., Lannering, B., Nordborg, C., Sabel, M., & Carén, H. (2015). MethPed: A DNA methylation classifier tool for the identification of pediatric brain tumor subtypes. *Clinical Epigenetics, 7*(1), 62. https://doi.org/10.1186/s13148-015-0103-3

Dash, P. K., Hochner, B., & Kandel, E. R. (1990). Injection of the cAMP-responsive element into the nucleus of Aplysia sensory neurons blocks long-term facilitation. *Nature, 345*(6277), 718–721. https://doi.org/10.1038/345718a0

Day, J. J., & Sweatt, J. D. (2011). Epigenetic modifications in neurons are essential for formation and storage of behavioral memory. *Neuropsychopharmacology, 36*(1), 357–358. https://doi.org/10.1038/npp.2010.125

Dusart, I., & Flamant, F. (2012). Profound morphological and functional changes of rodent Purkinje cells between the first and the second postnatal weeks: A metamorphosis? *Frontiers in Neuroanatomy, 6*, 11. https://doi.org/10.3389/fnana.2012.00011

Engel, A. K., Fries, P., König, P., Brecht, M., & Singer, W. (1999). Temporal binding, binocular rivalry, and consciousness. *Consciousness and Cognition, 8*(2), 128–151. https://doi.org/10.1006/ccog.1999.0389

Farhy-Tselnicker, I., & Allen, N. J. (2018). Astrocytes, neurons, synapses: A tripartite view on cortical circuit development. *Neural Development, 13*(1), 7. https://doi.org/10.1186/s13064-018-0104-y

Friston, K. J. (1997). Another neural code? *NeuroImage, 5*(3), 213–220. https://doi.org/10.1006/nimg.1997.0260

Georgopoulos, A. P., Schwartz, A. B., & Kettner, R. E. (1986). Neuronal population coding of movement direction. *Science, 233*(4771), 1416–1419. https://doi.org/10.1126/science.3749885

Gerhard, F., Haslinger, R., & Pipa, G. (2011). Applying the multivariate time-rescaling theorem to neural population models. *Neural Computation, 23*(6), 1452–1483. https://doi.org/10.1162/NECO_a_00126

Gerstner, W., Kreiter, A. K., Markram, H., & Herz, A. V. M. (1997). Neural codes: Firing rates and beyond. *Proceedings of the National Academy of Sciences, 94*(24), 12740–12741. https://doi.org/10.1073/pnas.94.24.12740

Goldin, M. A., Harrell, E. R., Estebanez, L., & Shulz, D. E. (2018). Rich spatio-temporal stimulus dynamics unveil sensory specialization in cortical area S2. *Nature Communications, 9*(1), Article 1. https://doi.org/10.1038/s41467-018-06585-4

Gräff, J., & Mansuy, I. M. (2008). Epigenetic codes in cognition and behaviour. *Behavioural Brain Research, 192*(1), 70–87. https://doi.org/10.1016/j.bbr.2008.01.021

Gray, C. M., & Singer, W. (1989). Stimulus-specific neuronal oscillations in orientation columns of cat visual cortex. *Proceedings of the National Academy of Sciences, 86*(5), 1698–1702. https://doi.org/10.1073/pnas.86.5.1698

Grover, S., Nguyen, J. A., & Reinhart, R. M. G. (2021). Synchronizing brain rhythms to improve cognition. *Annual Review of Medicine, 72*, 29–43. https://doi.org/10.1146/annurev-med-060619-022857

Guan, Z., Giustetto, M., Lomvardas, S., Kim, J.-H., Miniaci, M. C., Schwartz, J. H., Thanos, D., & Kandel, E. R. (2002). Integration of long-term-memory-related synaptic plasticity involves bidirectional regulation of gene expression and chromatin structure. *Cell, 111*(4), 483–493. https://doi.org/10.1016/s0092-8674(02)01074-7

Herrero-Navarro, Á., Puche-Aroca, L., Moreno-Juan, V., Sempere-Ferràndez, A., Espinosa, A., Susín, R., Torres-Masjoan, L., Leyva-Díaz, E., Karow, M., Figueres-Oñate, M., López-Mascaraque, L., López-Atalaya, J. P., Berninger, B., & López-Bendito, G. (2021). Astrocytes and neurons share region-specific transcriptional signatures that confer regional identity to neuronal reprogramming. *Science Advances, 7*(15), eabe8978. https://doi.org/10.1126/sciadv.abe8978

Isles, A. R. (2015). Neural and behavioral epigenetics; what it is, and what is hype. *Genes, Brain, and Behavior, 14*(1), 64–72. https://doi.org/10.1111/gbb.12184

Isles, A. R., & Wilkinson, L. S. (2008). Epigenetics: What is it and why is it important to mental disease? *British Medical Bulletin, 85*(1), 35–45. https://doi.org/10.1093/bmb/ldn004

Iyengar, S. (2003). The analysis of multiple neural spike trains. In *Advances on methodological and applied aspects of probability and statistics*. CRC Press.

Jarosiewicz, B., Chase, S. M., Fraser, G. W., Velliste, M., Kass, R. E., & Schwartz, A. B. (2008). Functional network reorganization during learning in a brain-computer interface paradigm. *Proceedings of the National Academy of Sciences, 105*(49), 19486–19491. https://doi.org/10.1073/pnas.0808113105

Jurjuţ, O. F., Nikolić, D., Singer, W., Yu, S., Havenith, M. N., & Mureşan, R. C. (2011). Timescales of multineuronal activity patterns reflect temporal structure of visual stimuli. *PLoS One, 6*(2), e16758. https://doi.org/10.1371/journal.pone.0016758

Kandel, E. R. (2000). *The molecular biology of memory storage: A dialogue between genes and synapses*. Nobel lecture. https://www.nobelprize.org/uploads/2018/06/kandel-lecture.pdf

Kasabov, N. K. (2018). *Time-space, spiking neural networks and brain-inspired artificial intelligence*. Springer Berlin Heidelberg.

Kim, H. S., & Sasaki, J. Y. (2014). Cultural neuroscience: Biology of the mind in cultural contexts. *Annual Review of Psychology, 65*, 487–514. https://doi.org/10.1146/annurev-psych-010213-115040

Kirmayer, L. J. (2011). The future of critical neuroscience. In *Critical neuroscience* (pp. 367–383). John Wiley & Sons, Ltd. https://doi.org/10.1002/9781444343359.ch18

Kirmayer, L. J., & Gómez-Carrillo, A. (2019). Agency, embodiment and enactment in psychosomatic theory and practice. *Medical Humanities, 45*(2), 169–182. https://doi.org/10.1136/medhum-2018-011618

Kitayama, S., & Park, J. (2010). Cultural neuroscience of the self: Understanding the social grounding of the brain. *Social Cognitive and Affective Neuroscience, 5*(2–3), 111–129. https://doi.org/10.1093/scan/nsq052

Kitayama, S., & Park, J. (2014). Error-related brain activity reveals self-centric motivation: Culture matters. *Journal of Experimental Psychology. General, 143*(1), 62–70. https://doi.org/10.1037/a0031696

Kitayama, S., & Tompson, S. (2010). Envisioning the future of cultural neuroscience. *Asian Journal of Social Psychology, 13*(2), 92–101. https://doi.org/10.1111/j.1467-839X.2010.01304.x

Kobayashi, C., Glover, G. H., & Temple, E. (2006). Cultural and linguistic influence on neural bases of "Theory of Mind": An fMRI study with Japanese bilinguals. *Brain and Language, 98*(2), 210–220. https://doi.org/10.1016/j.bandl.2006.04.013

Kobayashi, C., Glover, G. H., & Temple, E. (2007). Cultural and linguistic effects on neural bases of "Theory of Mind" in American and Japanese children. *Brain Research, 1164*, 95–107. https://doi.org/10.1016/j.brainres.2007.06.022

Koelsche, C., Schrimpf, D., Stichel, D., Sill, M., Sahm, F., Reuss, D. E., Blattner, M., Worst, B., Heilig, C. E., Beck, K., Horak, P., Kreutzfeldt, S., Paff, E., Stark, S., Johann, P., Selt, F., Ecker, J., Sturm, D., Pajtler, K. W., et al. (2021). Sarcoma classification by DNA methylation profiling. *Nature Communications, 12*(1), Article 1. https://doi.org/10.1038/s41467-020-20603-4

Kriegeskorte, N., & Golan, T. (2019). Neural network models and deep learning. *Current Biology, 29*(7), R231–R236. https://doi.org/10.1016/j.cub.2019.02.034

Krishna Temburni, M., & Jacob, M. H. (2001). New functions for glia in the brain. *Proceedings of the National Academy of Sciences, 98*(7), 3631–3632. https://doi.org/10.1073/pnas.081073198

Kundaje, A., Meuleman, W., Ernst, J., Bilenky, M., Yen, A., Heravi-Moussavi, A., Kheradpour, P., Zhang, Z., Wang, J., Ziller, M. J., Amin, V., Whitaker, J. W., Schultz, M. D., Ward, L. D., Sarkar, A., Quon, G., Sandstrom, R. S., Eaton, M. L., Wu, Y.-C., et al. (2015). Integrative analysis of 111 reference human epigenomes. *Nature, 518*(7539), Article 7539. https://doi.org/10.1038/nature14248

Kyrke-Smith, M., & Williams, J. M. (2018). Bridging synaptic and epigenetic maintenance mechanisms of the engram. *Frontiers in Molecular Neuroscience, 11*. https://www.frontiersin.org/articles/10.3389/fnmol.2018.00369

Landecker, H. (2005, January 11). Living differently in time: Plasticity, temporality and cellular biotechnologies. *Culture Machine.* https://culturemachine.net/biopolitics/living-differently-in-time/

Landecker, H. (2007). *Culturing life: How cells became technologies.* Harvard University Press. https://hdl.handle.net/2027/heb09113.0001.001

Leontiev, A. N. (1981). *Problems of the development of the mind.* Progress Publ.

Levenson, J. M., & Sweatt, J. D. (2005). Epigenetic mechanisms in memory formation. *Nature Reviews Neuroscience, 6*(2), Article 2. https://doi.org/10.1038/nrn1604

Levenson, J. M., Roth, T. L., Lubin, F. D., Miller, C. A., Huang, I.-C., Desai, P., Malone, L. M., & Sweatt, J. D. (2006). Evidence that DNA (cytosine-5) methyltransferase regulates synaptic plasticity in the hippocampus. *The Journal of Biological Chemistry, 281*(23), 15763–15773. https://doi.org/10.1074/jbc.M511767200

Li, S., Ye, Z., Mather, K. A., Nguyen, T. L., Dite, G. S., Armstrong, N. J., Wong, E. M., Thalamuthu, A., Giles, G. G., Craig, J. M., Saffery, R., Southey, M. C., Tan, Q., Sachdev, P. S., & Hopper, J. L. (2022). Early life affects late-life health through determining DNA methylation across the lifespan: A twin study. *eBioMedicine, 77.* https://doi.org/10.1016/j.ebiom.2022.103927

Lonze, B. E., & Ginty, D. D. (2002). Function and regulation of CREB family transcription factors in the nervous system. *Neuron, 35*(4), 605–623. https://doi.org/10.1016/S0896-6273(02)00828-0

Lux, V. (2018). Epigenetic programming effects of early life stress: A dual-activation hypothesis. *Current Genomics, 19*(8), 638–652. https://doi.org/10.2174/1389202919666180307151358

Malik, W. Q., & Ajemian, R. (2017). *Microarrays in the brain: Can they be used for brain-machine interface control?* (pp. 3–39). https://doi.org/10.1016/B978-0-12-800454-8.00001-X

Marshall, P., & Bredy, T. W. (2016). Cognitive neuroepigenetics: The next evolution in our understanding of the molecular mechanisms underlying learning and memory? *Npj Science of Learning, 1*(1), Article 1. https://doi.org/10.1038/npjscilearn.2016.14

Martins, J., Czamara, D., Sauer, S., Rex-Haffner, M., Dittrich, K., Dörr, P., de Punder, K., Overfeld, J., Knop, A., Dammering, F., Entringer, S., Winter, S. M., Buss, C., Heim, C., & Binder, E. B. (2021). Childhood adversity correlates with stable changes in DNA methylation trajectories in children and converges with epigenetic signatures of prenatal stress. *Neurobiology of Stress, 15,* 100336. https://doi.org/10.1016/j.ynstr.2021.100336

Mews, P., Calipari, E. S., Day, J., Lobo, M. K., Bredy, T., & Abel, T. (2021). From circuits to chromatin: The emerging role of epigenetics in mental health. *Journal of Neuroscience, 41*(5), 873–882. https://doi.org/10.1523/JNEUROSCI.1649-20.2020

Miller, C. A., & Sweatt, J. D. (2007). Covalent modification of DNA regulates memory formation. *Neuron, 53*(6), 857–869. https://doi.org/10.1016/j.neuron.2007.02.022

Miller, C. A., Campbell, S. L., & Sweatt, J. D. (2008). DNA methylation and histone acetylation work in concert to regulate memory formation and synaptic plasticity. *Neurobiology of Learning and Memory, 89*(4), 599–603. https://doi.org/10.1016/j.nlm.2007.07.016

Nicolelis, M. A. L., Ghazanfar, A. A., Faggin, B. M., Votaw, S., & Oliveira, L. M. O. (1997). Reconstructing the engram: Simultaneous, multisite, many single neuron recordings. *Neuron, 18*(4), 529–537. https://doi.org/10.1016/S0896-6273(00)80295-0

Paninski, L., Ahmadian, Y., Ferreira, D. G., Koyama, S., Rad, K. R., Vidne, M., Vogelstein, J., & Wu, W. (2010). A new look at state-space models for neural data. *Journal of Computational Neuroscience, 29*(0), 107–126. https://doi.org/10.1007/s10827-009-0179-x

Perez-Catalan, N. A., Doe, C. Q., & Ackerman, S. D. (2021). The role of astrocyte-mediated plasticity in neural circuit development and function. *Neural Development, 16*(1), 1. https://doi.org/10.1186/s13064-020-00151-9

Petersen, C. C. H. (2007). The functional organization of the barrel cortex. *Neuron, 56*(2), 339–355. https://doi.org/10.1016/j.neuron.2007.09.017

Pillow, J. W., Shlens, J., Paninski, L., Sher, A., Litke, A. M., Chichilnisky, E. J., & Simoncelli, E. P. (2008). Spatio-temporal correlations and visual signalling in a complete neuronal population. *Nature, 454*(7207), Article 7207. https://doi.org/10.1038/nature07140

Rahman, M. F., & McGowan, P. O. (2022). Cell-type-specific epigenetic effects of early life stress on the brain. *Translational Psychiatry, 12*, 326. https://doi.org/10.1038/s41398-022-02076-9

Rheinberger, H.-J. (1997). *Toward a history of epistemic things: Synthesizing proteins in the test tube*. Stanford University Press.

Rheinberger, H.-J. (2000). Beyond nature and culture: Modes of reasoning in the age of molecular biology and medicine. In A. Cambrosio, A. Young, & M. Lock (Eds.), *Living and working with the new medical technologies: Intersections of inquiry* (pp. 19–30). Cambridge University Press. https://doi.org/10.1017/CBO9780511621765.002

Rheinberger, H.-J. (2010). *An epistemology of the concrete: Twentieth-century histories of life*. Duke University Press. https://doi.org/10.2307/j.ctv11qdxmc

Ripoli, C. (2017). Engrampigenetics: Epigenetics of engram memory cells. *Behavioural Brain Research, 325*(Pt B), 297–302. https://doi.org/10.1016/j.bbr.2016.11.043

Rivera, C. M., & Ren, B. (2013). Mapping human epigenomes. *Cell, 155*(1). https://doi.org/10.1016/j.cell.2013.09.011

Santhanam, G., Ryu, S. I., Yu, B. M., Afshar, A., & Shenoy, K. V. (2006). A high-performance brain–computer interface. *Nature, 442*(7099), Article 7099. https://doi.org/10.1038/nature04968

Scott, C. A., Duryea, J. D., MacKay, H., Baker, M. S., Laritsky, E., Gunasekara, C. J., Coarfa, C., & Waterland, R. A. (2020). Identification of cell type-specific methylation signals in bulk whole genome bisulfite sequencing data. *Genome Biology, 21*(1), 156. https://doi.org/10.1186/s13059-020-02065-5

Seligman, R. (2017). "Bio-looping" and the psychophysiological in religious belief and practice: Mechanisms of embodiment in Candomblé trance and possession. In *The Palgrave handbook of biology and society* (pp. 417–439). Palgrave Macmillan. https://doi.org/10.1057/978-1-137-52879-7_18

Seligman, R., Choudhury, S., & Kirmayer, L. J. (2016). Locating culture in the brain and in the world: From social categories to the ecology of mind. In *The Oxford handbook of cultural neuroscience* (pp. 3–20). Oxford University Press.

Shadlen, M. N., & Movshon, J. A. (1999). Synchrony unbound: A critical evaluation of the temporal binding hypothesis. *Neuron, 24*(1), 67–77, 111–125. https://doi.org/10.1016/s0896-6273(00)80822-3

Singer, W. (1999). Neuronal synchrony. *Neuron, 24*(1), 49–65. https://doi.org/10.1016/S0896-6273(00)80821-1

Singer, W. (2021). Recurrent dynamics in the cerebral cortex: Integration of sensory evidence with stored knowledge. *Proceedings of the National Academy of Sciences, 118*(33), e2101043118. https://doi.org/10.1073/pnas.2101043118

Sohal, V. S., Zhang, F., Yizhar, O., & Deisseroth, K. (2009). Parvalbumin neurons and gamma rhythms enhance cortical circuit performance. *Nature, 459*(7247), 698–702. https://doi.org/10.1038/nature07991

Soler, L., Zwart, S. D., Israel-Jost, V., & Lynch, M. (Eds.). (2017). *Science after the practice turn in philosophy, history, and social studies of science (First issued in paperback)*. Routledge, Taylor and Francis Group.

Stogsdill, J. A., & Eroglu, C. (2017). The interplay between neurons and glia in synapse development and plasticity. *Current Opinion in Neurobiology, 42*, 1–8. https://doi.org/10.1016/j.conb.2016.09.016

Sui, J., Liu, C. H., & Han, S. (2009). Cultural difference in neural mechanisms of self-recognition. *Social Neuroscience, 4*(5), 402–411. https://doi.org/10.1080/17470910802674825

Sweatt, J. D. (2013). The emerging field of neuroepigenetics. *Neuron, 80*(3), 624–632. https://doi.org/10.1016/j.neuron.2013.10.023

Tang, Y., Zhang, W., Chen, K., Feng, S., Ji, Y., Shen, J., Reiman, E. M., & Liu, Y. (2006). Arithmetic processing in the brain shaped by cultures. *Proceedings of the National Academy of Sciences, 103*(28), 10775–10780. https://doi.org/10.1073/pnas.0604416103

Truccolo, W., Eden, U. T., Fellows, M. R., Donoghue, J. P., & Brown, E. N. (2005). A point process framework for relating neural spiking activity to spiking history, neural ensemble, and extrinsic covariate effects. *Journal of Neurophysiology, 93*(2), 1074–1089. https://doi.org/10.1152/jn.00697.2004

Uchida, S., Teubner, B. J. W., Hevi, C., Hara, K., Kobayashi, A., Dave, R. M., Shintaku, T., Jaikhan, P., Yamagata, H., Suzuki, T., Watanabe, Y., Zakharenko, S. S., & Shumyatsky, G. P. (2017). CRTC1 nuclear translocation following learning modulates memory strength via exchange of chromatin remodeling complexes on the Fgf1 gene. *Cell Reports, 18*(2), 352–366. https://doi.org/10.1016/j.celrep.2016.12.052

Uhlhaas, P., Pipa, G., Lima, B., Melloni, L., Neuenschwander, S., Nikolić, D., & Singer, W. (2009). Neural synchrony in cortical networks: History, concept and current status. *Frontiers in Integrative Neuroscience, 3*. https://www.frontiersin.org/articles/10.3389/neuro.07.017.2009

VanLeeuwen, J.-E., Rafalovich, I., Sellers, K., Jones, K. A., Griffith, T. N., Huda, R., Miller, R. J., Srivastava, D. P., & Penzes, P. (2014). Coordinated nuclear and synaptic shuttling of afadin promotes spine plasticity and histone modifications. *The Journal of Biological Chemistry, 289*(15), 10831–10842. https://doi.org/10.1074/jbc.M113.536391

von der Malsburg, C. (1994). The correlation theory of brain function. In E. Domany, J. L. van Hemmen, & K. Schulten (Eds.), *Models of neural networks: Temporal aspects of coding and information processing in biological systems* (pp. 95–119). Springer. https://doi.org/10.1007/978-1-4612-4320-5_2

Wang, B., & Dudko, O. K. (2021). A theory of synaptic transmission. *eLife, 10*, e73585. https://doi.org/10.7554/eLife.73585

Wang, T., Morency, D. T., Harris, N., & Davis, G. W. (2020). Epigenetic signaling in glia controls presynaptic homeostatic plasticity. *Neuron, 105*(3), 491–505.e3. https://doi.org/10.1016/j.neuron.2019.10.041

Watanabe, S. (2015). Slow or fast? A tale of synaptic vesicle recycling. *Science, 350*(6256), 46–47. https://doi.org/10.1126/science.aad2996

Wei, J., Xiong, Z., Lee, J. B., Cheng, J., Duffney, L. J., Matas, E., & Yan, Z. (2016). Histone modification of Nedd4 ubiquitin ligase controls the loss of AMPA receptors and cognitive impairment induced by repeated stress. *The Journal of Neuroscience: The Official Journal of the Society for Neuroscience, 36*(7), 2119–2130. https://doi.org/10.1523/JNEUROSCI.3056-15.2016

Wytock, T. P., & Motter, A. E. (2020). Distinguishing cell phenotype using cell epigenotype. *Science Advances, 6*(12), eaax7798. https://doi.org/10.1126/sciadv.aax7798

Zarrinkoob, L., Ambarki, K., Wåhlin, A., Birgander, R., Eklund, A., & Malm, J. (2015). Blood flow distribution in cerebral arteries. *Journal of Cerebral Blood Flow & Metabolism, 35*(4), 648–654. https://doi.org/10.1038/jcbfm.2014.241

Zhang, X., & Ho, S.-M. (2011). Epigenetics meets endocrinology. *Journal of Molecular Endocrinology, 46*(1), R11–R32.

Zhu, T., Liu, J., Beck, S., Pan, S., Capper, D., Lechner, M., Thirlwell, C., Breeze, C. E., & Teschendorff, A. E. (2022). A pan-tissue DNA methylation atlas enables in silico decomposition of human tissue methylomes at cell-type resolution. *Nature Methods, 19*(3), Article 3. https://doi.org/10.1038/s41592-022-01412-7

Chapter 5
The Molecular Epigenetic Lens

> Human brains and even the most powerful computers have difficulty in handling and making sense of the overwhelming flow of data generated by recent high-throughput technologies. This was easier when low throughput, more integrative methods based on biochemistry and microscopy dominated biological research. Nowadays, the need for organising concepts is ever more important, otherwise the mass of available data can generate only "building ruins"—the bricks without an architect. (Scherrer, 2018, p. 1)

A lens sharpens the image and often works as a magnifying glass. By placing the single neuron under the molecular epigenetic lens, we intend to gather information about the inner molecular workings of this exciting cell. However, obtaining a clear picture is difficult. Neuroepigenetic data do not reflect the epigenome of a single neuron, at least in most studies. The epigenetic markers of interest—whole genome or candidate gene-related DNA methylation patterns, specific histone modifications, a specific set of microRNAs, or the presence and quantity of epigenetic active enzymes—are commonly measured in whole brain tissue. Whole brain tissue includes all cell types present in the brain. Sometimes, tissue samples are drawn from a specific brain area, sometimes not. Human brain tissue is rare and, if not sampled post-mortem, is often retrieved from scrap tissue during brain surgery. Alternatively, cell cultures or organoids[1] are used, with the limitation that the in vitro conditions lead to specific developmental pathways of these neurons, which are also reflected in the epigenome (Amiri et al., 2018; Vieira et al., 2019; see discussion in the previous chapter). Methods targeting specific cell types—neurons, even specific subtypes of neurons, or glia cells—are, in principle, available (Bheda & Schneider, 2014; Hayashi-Takanaka et al., 2020; Karemaker & Vermeulen, 2018); however, their work intensity and costs strongly limit their wider use. To measure the epigenome within a single neuron, the neuron must first be identified and removed from the brain tissue. This is usually done either by fluorescence-activated cell sorting (FACS), laser capture microdissection (LCM) from frozen brain slices, or combined

---

[1] Organoids are in vitro grown organs or parts of organs which are used as experimental models. For brain tissue, the first organoids were developed to model microencephaly (Lancaster et al., 2013), and are now explored for wider use in human brain research (Agboola et al., 2021; Di Lullo & Kriegstein, 2017; Kadoshima et al., 2013; Qian et al., 2016).

V. Lux, *The Neuron in Context*, SpringerBriefs in Psychology,
https://doi.org/10.1007/978-3-031-55229-8_5

methods (Backes & Hemby, 2003; Hempel et al., 2007; Liu, 2010; Lobo et al., 2006). In standard protocols, these methods often result in cell loss, and not all cells surviving the procedure will retain their neurites, so that only the cell nuclei can be used (Hempel et al., 2007; for an example, see Södersten et al., 2018). These procedures work well for the study of chromatin-bound epigenetic mechanisms, such as DNA methylation patterns and histone modifications, as well as for the quantification of epigenetic active enzymes. The case is slightly more difficult for the mechanisms of RNA interference involved in synaptic processes. Here, changes in miRNAs participating in the regulation of synaptic activity are more present directly at the synapses and in axons and dendrites than in the cell body (for a comprehensive review, see Olde Loohuis et al., 2012). Although it is, in principle, possible to collect synapses from brain tissue, the costs and work intensity do not match the possible gain of information at the current state of our knowledge in most research settings. Studies on miRNAs involved in synaptic activity are often limited to the presence of these miRNAs in the cell body, or they depend on the use of cell cultures and knock-out model organisms (Ruberti et al., 2012). However, the latter requires that the miRNAs of interest be known.

When considering the highly individualized developmental and activation history of a single neuron, the single-cell approach seems to be the way to go. Accordingly, some researchers called for such a shift in methodology (Bheda & Schneider, 2014; Clark et al., 2016; Karemaker & Vermeulen, 2018; O'Neill et al., 2022; Walter & Schickl, 2019). This approach implies detection based on very limited molecular material, which only a few laboratories have mastered. As an alternative, most studies use broad measurements of DNA methylation, histone modifications, and epigenetically active enzymes across the whole brain tissue. An intermediate step is the use of cell type-specific measurements of epigenetic markers from pooled cells. These are also unable to fully account for the individual developmental history of single neurons, as they represent a population estimate of the pooled cells. Thus, epigenetic data are usually collected across several neurons, subtypes of neurons, and other groups of brain cells. In addition, most detection methods for epigenetic marks have only limited coverage across the genome, and attempts to analyze all known mechanisms and modifications in a unique cell over its dynamic course of development are still facing tremendous challenges (Law & Holland, 2019). New methods with more precise coverage are constantly being developed (e.g., Hayashi-Takanaka et al., 2020; Heiss et al., 2019; Khodadadi et al., 2021; Laszlo et al., 2013; Önder et al., 2015; Rand et al., 2017), but the full epigenome detection of an individual cell will still be out of reach for quite some time. This adds additional blurriness to epigenetic information, even at the single-cell level. The question arises as to what type of information this data provides about the neuron in its context and how, if at all, can we relate this information or knowledge to psychological function. What is it that we can see through the molecular epigenetic lens?

Epigenetic Layers Within Multicellular Tissue

When measuring epigenetic markers across cells, differences between these cells are no longer detectable. We assume that these are equally distributed among the cells in the sample. This is unproblematic in relatively homogenous tissue, and it is also less significant when epigenetic modifications within the target cells are different enough to stand out against the epigenetic background noise of the cell mixture, as has been shown for some whole-genome DNA methylation patterns in cancer cells (Jones & Baylin, 2007).[2] In contrast, in heterogeneous cell samples, such as whole blood or brain tissue, the cell composition becomes an issue for any detection strategy. Here, differences between two measurement time points or treatment groups could just indicate differences in cell composition (Jaffe & Irizarry, 2014; Liu et al., 2013; Reinius et al., 2012; Teschendorff et al., 2015; Teschendorff & Zheng, 2017). These differences can result from incidental variation during sample collection, natural variation in the distribution of these cells across the tissue and at the sample sites, or variability in cell composition between individuals. In some cases, the latter may result from long- or short-term disease-related differences in cell composition and development, such as changes in white blood cell counts in relation to inflammatory processes within the body, and thus may be informative on their own and useful as a biomarker (Liu et al., 2013). In some cases, differences in cell composition are causal to the phenotypic differences under study.[3] However, if not addressed, this issue remains a primary source of false-positive or false-negative findings.

This has been widely discussed in epigenome-wide association studies (EWAS) of DNA methylation using blood cells (Bearer & Mulligan, 2018; Kumsta, 2019; Lowe & Rakyan, 2014), and experimental protocols that account for cell-composition bias have been developed. Methods range from a variety of statistical correction methods to the empirical determination of cell type distributions in the individual samples under study (Brägelmann & Lorenzo Bermejo, 2018; Dieckmann et al., 2022; McGregor et al., 2016; Rahmani et al., 2019; Teschendorff & Zheng, 2017). Estimate-based strategies are less work-intensive and, due to their precision across datasets, have become the gold standard (Qi & Teschendorff, 2022). They are either reference-based, using known cell-specific gene expression or DNA methylation patterns as reference measures to estimate the cell composition of a specific sample, or reference-free, estimating the cell composition by identifying latent variance factors across samples. The estimates for individual cell types are then used as control variables when analyzing group differences in DNA methylation, histone modifications, or microRNA distribution.

---

[2] The aggregation level of these measurements is comparable to those of standard EEG curves, measured at the skull, which are interpreted as the summed neural activity of larger groups of cells.

[3] For a conceptual discussion on how to interpret epigenome-wide association studies in this context, see Lappalainen and Greally (2017).

In light of these methodological debates, researchers have turned to articulating more precisely what the measurements entail. For example, Rhamani et al. explicitly use the wording "tissue-level bulk methylation" (2019, p. 1) when referencing DNA methylation patterns across different cells. This labeling clearly improves science communication. From the perspective of the old neuron theory, it also gives rise to the impression that these bulk-tissue epigenetic measurements are not precise enough to be informative and that measurements at the single-neuron level are to be favored. Tissue-level bulk data are clearly limited in terms of information about the epigenome of a single neuron. However, if we assume that the cell heterogeneity of brain tissue is potentially related to functional specialization, as the differences in cell composition between functionally different brain areas suggest, tissue-level bulk methylation is a desirable measure on its own. Whether single-cell or tissue-level epigenetic modifications are more informative depends on the phenotype of interest in a specific study (Lappalainen & Greally, 2017). This also means that propagating single-cell measurements as the future gold standard for neuroepigenetic studies is presumptuous at this stage, and the required specification level of the data clearly depends on the research question. Single-cell measurements are the data of choice when characterizing the neuron as a specific cell type, in contrast to other somatic cells. At this single neuron level, differences between individual neurons are more strongly related to homeostasis, as maintaining homeostasis is vital to the cell's survival and normal functioning. As discussed in the previous chapter, the patterns of neural functioning of a single neuron are highly individual and dynamic due to their context dependence and their role in a neuron's functional specialization. The limits that this dynamic individualization induces on efforts to match the single neuron activity to neural activity on a larger scale, as discussed, question the role of the single neuron as a basic unit of a specific behavior or psychological function. When we study how this individualization is reflected in its epigenetic cell state, we need to consider that functionally relevant changes are not easily detectable against homeostasis-related modifications at the single-neuron level. Instead, functionally relevant changes as they relate to changes in higher-order neural activity and psychological functions will often be detectable and actually become functionally relevant at a certain threshold of neurons or even at the bulk-tissue level, adding, for example, modifications in glia cells. The epigenetic data do not represent the epigenetic state of single cells but some aggregated measurements for the respective cell ensemble or tissue at this specific sampling site. In this regard, epigenetic data retrieved from bulk tissue contains multiple layers of information.

This introduces some fuzziness into these measurements, going against a causal mechanistic match between the molecular markup of a single cell and neural activity at the network or even higher functional levels. However, this fuzziness is not due only to imprecise measurements because of technological constraints. It also mimics the relationship between the molecular state of the single neuron and its neural function at the population level. Our difficulties in enduring this fuzziness reflect the fact that it stands in contrast to our traditional view of the neuron as a basic unit of neural function. However, epigenetic information at the bulk tissue level provides a sensitive marker for the local epigenetic state of a specific tissue sample. Simultaneously, the bulk-tissue epigenetic state also provides a potential biomarker for the specific molecular context of individual neurons within a sample. Thus, although bulk-tissue-level epigenetic data do not describe the molecular state of the single cell, they still contribute to our ability to estimate, for example, the likelihood of single neurons in a specific sample to successfully participate in neural activity, the readiness of neurons to engage in new network connections, the likelihood of specific functional failures of neurons, or their potential synaptic plasticity range. Most importantly, the bulk-tissue level measurement provides a rough marker for the epigenetic state related to the local network level, for example, small neural populations exhibiting activation loops. For both uses, further characterization of such bulk-tissue-level epigenetic markers will eventually differentiate between cellular (homeostasis-related) changes, stochastic background noise, and functionally relevant modifications. Cell type-specific epigenetic profiles and cell composition estimates provide the first markers for this characterization, but further measurements unraveling the layers of epigenetic information at the bulk tissue level are needed.

The Time-Dependence of Epigenetic Markers

Another source of the multilayer character of epigenetic information is its time dependence. Age-related differences in DNA methylation in differentiated somatic cells were among the first indications that epigenetic marks change over the lifespan, even after cell differentiation is completed (Fraga et al., 2005). However, in most cases, the timing of epigenetic modifications and the cellular feedback loops in which they are involved are still not entirely understood. For epigenetic modifications in brain cells, we only start to understand at which stages of cell differentiation and development these are established, and what the timeline of the transcription cycles establishing them looks like (Hayashi-Takanaka et al., 2020; Södersten et al., 2018; Wang et al., 2016). We do not know exactly how quickly DNA methylation at specific CpG sites in neurons or glia cells can change in response to an environmental signal via DNA methylation, demethylation, or remethylation. We also do not know how strong such a signal needs to be to induce these processes or how long it needs to be preserved. The same is true for histone modifications, although the general assumption and experimental observations are that they operate on much shorter time frames (Hayashi-Takanaka et al., 2020; Niu et al., 2015; Riffo-Campos et al., 2015).

Although measurement models and experimental designs often assume that epigenetic changes over time follow a linear timeline, the initial findings indicate a more complex chronological structure of epigenetic information, at least for DNA methylation patterns (Oh & Petronis, 2021). DNA methylation is involved in and responsive to seasonal and even daily regulation of gene expression (Azzi et al., 2014; Liang et al., 2019; Viitaniemi et al., 2019; Xia et al., 2015), and variability in these dynamics has been shown to be partially related and even predictive of aging and disease onset (Oh et al., 2018, 2019). Similar to variations in cell type composition, variations in DNA methylation variability have been reported to be associated with complex disease phenotypes, such as rheumatoid arthritis (Webster et al., 2018) and type 1 diabetes (Paul et al., 2016). Thus, we need to take into account that epigenetic information consists of several chronological layers, some of them following linear change over time, while others represent shorter (minutely, daily) or longer (seasonal) cyclical changes; some of them are stochastic, and some of them signal-induced, some of them without specific function, and some of them functionally related to phenotypic differences.

This issue complicates further, considering that in addition to stochastic or activity-induced changes and their potential variability across the genome, there are developmental periods during which there is a higher degree of permeability for epigenetic modifications and their long-term conservation in a tissue- and locus-dependent manner (Bock et al., 2015; Martins et al., 2021; Rulands et al., 2018). Epigenetic modifications, especially chromatin-bound histone modifications and DNA methylation patterns, often serve developmental functions in cell, tissue, and organ development. The precise characterization of critical and sensible periods for loci-specific epigenetic modifications and their stabilization would help to further unravel the different layers of epigenetic information. Oh and Petronis (2021) strongly emphasized this point, arguing that the temporal dimension is a key component that epigenetics adds to the study of development and disease. Accordingly, they call for new methodological approaches that, in contrast to traditional cross-sectional designs, are able to capture "dynamic epigenomes" (Oh & Petronis, 2021, p. 533). However, to fulfill the promise of this new "chrono-epigenetic perspective," as they coined it, the potentially different developmental and homeostasis-related timelines of epigenetic modifications need to be identified for the specific types of target tissue. In the case of neuroepigenetics, this includes potential differences between brain areas and brain cell types, such as glia cells, neurons, and their subtypes, but also the developmental trajectories of different peripheral tissues often used as surrogates in clinical studies, such as whole blood and cord blood, subgroups of blood cells, buccal cells, and saliva—to name just the most commonly used. Knowing the timely coordination of epigenetic modifications in different tissue types is an important prerequisite for effectively tracing them in an experimental context or a clinical study. Otherwise, measurements taken at different time points during such a daily or seasonal cycle could easily result in false-positive or false-negative findings, comparable to the confounding effects of cell-type heterogeneity (Oh & Petronis, 2021). Accordingly, we need to consider that different research approaches are not only limited in their accessibility to the target tissue but also in the timeframe they are able to cover (see Fig. 5.1).

To entangle the different chronological layers, reference measures for different chronological dynamics in different tissues, such as daily or seasonal variability scores, are needed. One group of such measures, which already provide some insight into time-dependent changes of the epigenome and serve as potential epigenetic markers, are the so called "epigenetic clocks" (Horvath, 2013, p. 1). Epigenetic clocks have been developed to predict chronological age using the DNA

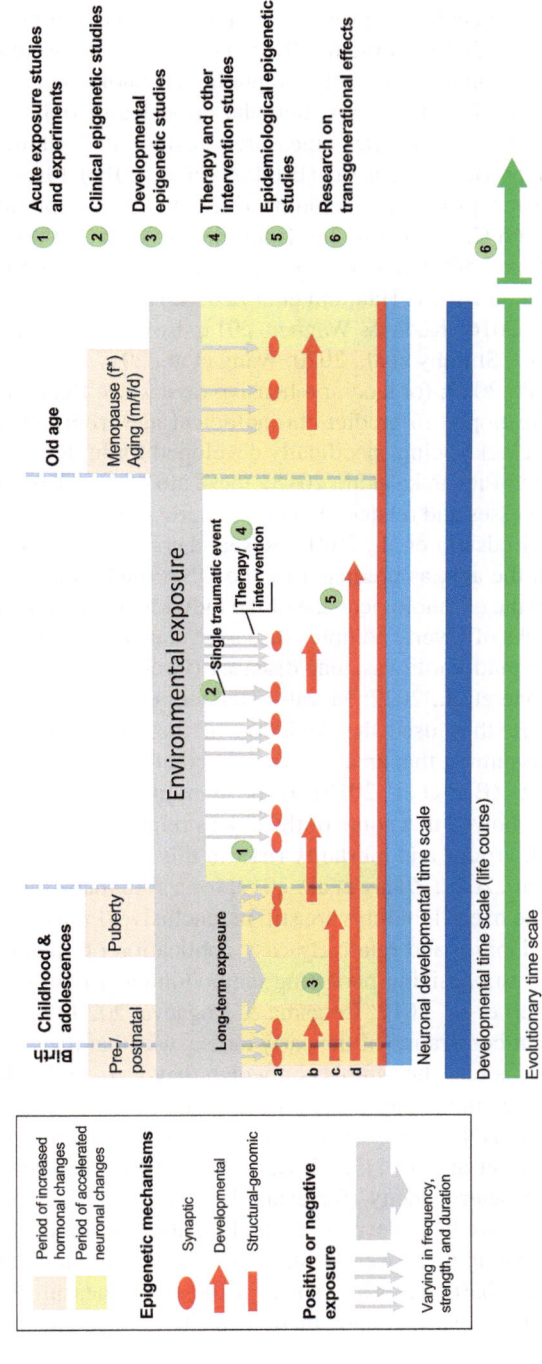

Fig. 5.1 Timeframes of epigenetic mechanisms across the lifespan and related research approaches

methylation status of a set of single CpG sites across the genome. They are based on the assumption that some of the stochastic changes in the DNA methylome are related to aging due to a decline in proper genomic and structural-epigenetic maintenance (Hannum et al., 2013; Horvath, 2013). The first clocks were developed by Horvath (2013) and Hannum et al. (2013) to predict chronological age from whole blood methylomes based on the DNA methylation states of only 71 and 353 of approximately 480,000 CpG sites measured using the Illumina Infinium HumanMethylation450 BeadChip assay (Bibikova et al., 2011). Meanwhile, several other clocks have been proposed, including different clocks for different tissue types and species, with CpG sets ranging from three to over 20,000, most of them ranging between 35 and 500 CpG sites (Bocklandt et al., 2011; de Lima Camillo et al., 2022; Galkin et al., 2021; Hannum et al., 2013; Horvath, 2013; Huang et al., 2015; Knight et al., 2016; Koch & Wagner, 2011; Levine et al., 2018; Li et al., 2022b; Lu et al., 2019; Shireby et al., 2020; Wang et al., 2017; Weidner et al., 2014; Zbieć-Piekarska et al., 2015; for a comprehensive review see Bergsma & Rogaeva, 2020). Originally developed to predict chronological age from DNA methylation patterns, with some clocks being specifically developed for high-precision forensic age predictions (Zbieć-Piekarska et al., 2015), these clocks are also used to identify biological aging processes and related phenotypes across different tissues (Bergsma & Rogaeva, 2020; Grodstein et al., 2021). Researchers have begun to explore the differences between the age, as predicted by the DNA methylation patterns of the epigenetic clock and the chronological age of the individual, and to identify associations with disease risks of several complex diseases related to aging, such as cancer, Alzheimer's disease, and cardiovascular diseases (Grodstein et al., 2021; Huang et al., 2019; Johnstone et al., 2022; Morales Berstein et al., 2022). As a basis for these risk predictions, they used the deviation of the epigenetic clock from the chronological age, assuming that an accelerated predicted age indicates higher disease risk and mortality (Bell et al., 2019). These attempts were moderately successful. The first studies showed that some of the clocks reliably differentiated between different disease risk groups and mortality risks in this regard (Huang et al., 2019; Morales Berstein et al., 2022; Zhang et al., 2019; for a systematic review, see Oblak et al., 2021), while the overall findings are still inconclusive (Fransquet et al., 2019).

Following the reliability and relatively easy application of epigenetic clocks, a debate has emerged, emphasizing persisting methodological issues with the wider use of the clocks (Bell et al., 2019; Bergsma & Rogaeva, 2020). For example, one of the methodological constraints of the early clocks was that they did not account for cell heterogeneity across the samples on which they were built. Hence, it is not clear whether the DNA methylation patterns associated with aging would reflect age-related changes in DNA methylation levels or age-related changes in cell type composition (Hannum et al., 2013). Although some of the later-developed clocks account for cell-type heterogeneity (for example, Zhang et al., 2019) and several tissue-specific clocks have been developed, the functional relationship between the clock measures and the aging process is still under investigation (Bell et al., 2019; Bergsma & Rogaeva, 2020). The CpG sites for the clock measures were selected based on statistical properties and their contribution to the prediction of chronological age. Consequently, only a few sites used to calculate the clocks are associated with known processes of cellular aging or aging in general, or with aging and cancer-related genes,[4] leaving a blind spot on their potential mechanistic role in aging (Bell et al., 2019; Bergsma & Rogaeva, 2020; Li et al., 2022a). Attempts to extrapolate a mechanistic role for the aging process or the development of age-related diseases failed in most cases. Thus, as originally assumed, most of the differences in DNA methylation at these CpG sites likely represent stochastic modifications over time, which is supported by the observation that several epigenetic clocks with different sets of CpG sites can be used to reliably predict chronological age. To overcome this constraint, specialized clocks associated with cell aging mechanisms, such as those that measure mitotic age based on cell division count estimates, have been developed and explored (Bell et al., 2019; Yang et al., 2016). Further methodological constraints of the current epigenetic clocks include reduced reliability and validity in diverse populations and minorities, which are typically not included in the construction of the clocks, and potential sex differences, which are still underexplored (Bell et al., 2019).

Nevertheless, epigenetic clocks and their wider use in epigenetic studies are a first step in unraveling time-dependent layers of epigenetic information. They can be used as covariates in EWAS and longitudinal epigenetic studies to control for age-related variations in DNA methylation patterns across individuals, thereby reducing the noise that often masks the functional relationships of interest (Bell et al., 2019). Thus, epigenetic clocks are already used as a type of information filter at the bulk tissue level for the time dependence of epigenetic modifications. The different clocks help explore tissue-specific patterns of age-related DNA methylation changes and contribute to the characterization of time-dependent layers of epigenetic information. They allow the determination of tissue-specific sensitive periods for stochastic DNA methylation modifications across the genome, for example, characterized by a faster pace of biological aging in certain developmental periods compared to other periods across the lifespan. First studies already reported those differences in the pace of epigenetic clocks during typical developmental periods, such as early childhood and adolescence (Simpkin et al., 2016, 2017).

While constituting most of its practical use, for example, an easy inclusion in other biostatistical analyses as covariates, the aggregated nature of the epigenetic clock scores also implies a tremendous loss of information regarding the underlying genomic structure of these DNA methylation patterns. The same would be true for comparable scores of daily or seasonal variability of epigenetic marks. Thus, the question arises whether other approaches would allow for a more detailed account of the epigenome's information structure at the multicellular or bulk tissue level.

---

[4] Some of the few examples are CpG sites related to *elovl2* (elongation of very long-chain fatty acids protein 2) and *EDARadd* (EDAR associated via death domain).

The Interdependence of Genetic and Epigenetic Information

One way to further characterize the information structure of the epigenome is to use its dependence on the DNA sequence and, thus, parts of the related genomic information. The fact that epigenetic information depends on genomic information is obvious for RNA interference, as miRNAs and other non-coding RNAs are the (by) products of transcription loops and depend directly on the underlying DNA sequence. To some extent, this also applies to histone modifications, as the genetic markup and related transcription loops also affect the 3-D structure of chromatin (Gorkin et al., 2019). As a result, some histones are more accessible and prone to specific modifications than others. It seems that a combination of genetic sequence-dependent and sequence-independent mechanisms interact in the establishment and, most importantly, the maintenance of histone modifications during cell differentiation (Wang & Moazed, 2017). The link is complex, involving several feedback loops, and extrapolating the likelihood of specific histone modifications from the DNA sequence is not straightforward.

This is clearly different for DNA methylation. Due to the biophysical conditions of the DNA, to our current knowledge, methyl groups bind primarily at CpG sites within the DNA strand, and the distribution of those CpG sites is not random across the genome. The interpretation of DNA methylation as a gene-regulating mechanism stems to a huge part from this non-random distribution. Specifically, at the very beginning of molecular epigenetics, the cumulation of CpG sites in CpG islands and the overrepresentation of such islands within the promoter regions of genes, prior to transcription start sights, gave rise to the functional interpretation of DNA methylation as a dynamic mechanism regulating gene expression like a power switch. Meanwhile, complex patterns of CpG distributions across the human genome have been described (Laurent et al., 2010; Lister et al., 2009), and their functional relevance is further explored. In addition, DNA methylation at sites other than CpG sites, such as CpA, CpT, and CpC, is intensively studied (for review, see Jang et al., 2017). In the human genome, non-CpG methylation is estimated to comprise 0.02% of the total methylome in differentiated somatic cells, whereas it is estimated to be present in larger amounts in embryonic stem cells and brain tissue (Laurent et al., 2010; Lister et al., 2009). Although still marginal compared to DNA methylation at CpG sites, the higher presence of non-CpG methylation in embryonic stem cells and brain cells compared to other somatic cells indicates a more dynamic DNA methylome and a potential functional role of non-CpG methylation in these cell types. Finally, genetic variation in DNA sequences encoding epigenetic modifying enzymes such as DNA methyltransferases or histone deacetylases also affects the structure and maintenance of the epigenome. Thus, genetic information strongly contributes to the characterization of the epigenetic landscape over the lifespan of a neuron.

Solving Parts of the Puzzle with Information Theory: Potential Pitfalls

For both, genetic and epigenetic information, information theory has been proposed as a productive framework to characterize its information structure (Jenkinson et al., 2017, 2018, 2019; Jost, 2020; Jost & Scherrer, 2014; Scherrer & Jost, 2007). Information theory has the potential to capture both levels and their functional dynamics. For example, it allows differentiation between the regulatory and coding parts of cellular information and accounts for the temporal and topological dimensions of the information structure. In addition, the framework allows for the inclusion of higher regulatory levels, such as hormonal influences and synaptic activity, thereby providing conjunctions for a translation to the neural network level. Nonetheless, the application of information theory to the analysis of neuroepigenetic data and the integration of genetic and epigenetic information within this framework faces some challenges.

The formalized mathematical version of information theory referred to for the modeling of genetic and epigenetic data was first introduced by Shannon (1948). He proposed understanding information as something that is exchanged between a sender and a receiver to reduce variation within a random set of potential states of a system so that the prediction of a specific state is more likely. Successfully employed in the context of automated coding and decoding processes within computer sciences, the generalization of Shannon's mathematical concept of information to all kinds of communication and information exchanges has always been heavily disputed. In the context of psychological science, the use of Shannon's information theory has been controversial (Cronbach, 1955; Luce, 2003; Sayood, 2018). Attempts to model psychological processes, including cognition, memory, and language processing, based on this mathematically formalized concept of information have been repeatedly refuted (Cronbach, 1955; Laming, 2001; Luce, 2003). The abstract notion and formula of information as reduced randomness clearly miss the qualitative dimension and multidimensionality of human communication and meaning production, which are key to human experiences and communication acts. More complex information structures, such as multi-, non-matching, or contradictory information, are represented as failed communication within this formalized system, although they are an integral part of human communication (Luce, 2003). Even for the link between cognition and the neural level, computational models based on the notion of encoded information and developed to bridge, for example, cognition and memory with neural activity represented by spike trains, have been criticized as misleading, too simple, and improper use of computer metaphors (Brette, 2019).

In neuroscience, the debate has evolved in a different way. Here, different variants[5] of information theory have been applied to different layers of neural

---

[5] For a discussion of the theoretical differences between those variants as well as mathematically formalized accounts of Information Theory and the Integrated Information Theory as proposed by Tononi and Koch (Tononi, 2004; Tononi et al., 2016), which tries to integrate some aspects of

information to detect patterns within neural data streams (Borst & Theunissen, 1999; de Ruyter van Steveninck et al., 1997; MacKay & McCulloch, 1952; Timme & Lapish, 2018; Tononi, 2004; Tononi et al., 2016; Winograd & Cowan, 1963, for a review of information theory's theoretical development and application in neuroscience, see, for example, Dimitrov et al., 2011). Nonetheless, even among researchers applying information theory to neural data, its correspondence to the underlying biological processes is heavily debated (Jost, 2021; Koch, 1996, 2019). Neural activity within the brain and its relationship to psychological functions depend on the brain's material structure and how this structure–function relationship evolves over time—or, in other words, consciousness is embodied (Wilson, 2002). As we have seen, the single neuron and its function are intertwined with this materiality of the brain, resulting in a complex co-dependence up to the point where the single neuron can no longer be understood as the ultimate unit of neural activity. Thus, abstracting from this material basis in calculating event probabilities based only on the possible variants of the information sequence of the spike trains of firing neurons may not adequately reflect the biological processes underlying neural function, which heavily constrain and define the information structure (Koch, 2019). Even if we abstract from the wider cellular and network context and consider only the connection between two neurons, the physical preconditions in the "sending" as well as in the "receiving" neuron necessary for the transmission of an action potential are not easily described with the formalized, reproducible format required by information theory, due to, for example, their many-to-many and potentially cyclic interrelationship. In addition, oversimplification of regulatory mechanisms and context factors, such as treating action potentials and firing rates within and across neurons as independent while they are hierarchically intertwined, and the inability to account for multiple parallel coding mechanisms have been raised as potential limitations for the application of information theory to neural function (Brette, 2019). At the level of small neuron populations, Schöner et al. emphasized that the recurrent nature of neural activity within these networks and the implied neural dynamics "are not conceptually compatible with the information theoretical notions of encoding. [...] The output of any neuron depends not only on the inputs to the network but also on the current state of activation in the network, which reflects the recent history of activation and stimulation. Different histories of stimulation leading up to the same instantaneous stimulus lead to different activation patterns" (Schöner et al., 2015, p. 10). Jost, who demonstrated the use of information theory for the analysis of genetic data and further developed concepts for this purpose, also argues for caution. As he points out, in the context of biological systems and organism-environment interactions, it is often unclear which uncertainty is reduced by the signal that is supposed to transmit or encode the information (Jost, 2020). He further emphasized that the knowledge that the receiving system possesses to decode the signal, may it stem from its memory, some external context, or some regulatory scheme, needs to be carefully specified to successfully apply information

---

experienced consciousness from a phenomenological perspective with information theory, see, for example, Dimitrov et al. (2011), Jost (2021), and Brette (2019).

theoretical concepts to biological processes. Thus, although information theory is currently applied in neuroscience to identify patterns within the complex stream of neural measurements, the extent to which these patterns reflect the underlying biological processes is still debated.

In opposition to the mismatch of the sender-receiver model with the observed biological and psychosocial processes, other approaches have been used to model signal transmission across multiple biological, psychosocial, and cultural levels, some of which even predate information theory. These include, among others, Jakob von Uexküll's Umwelt-theory (von Uexküll, 1926, 2010, 1992/1984), Karl Bühler's Organon Model (Bühler, 1960, 1999), Jesper Hoffmeyer's Biosemiotics (Hoffmeyer, 2008), and Susan Oyama's Ontogeny of Information (Oyama, 2000). With these approaches, signal transmission in biological systems is conceptualized within the context of organism-environment interactions, accounting for developmental affordances, intentionality, subjective experiences, and, at the level of psychological functions and human communication, the social-cultural embeddedness of language and meaning production.

The question that arises in this context is whether it is adequate to use the same model for signal transmission and information processing across all levels of a living system. Here, the distinction between the animate and the inanimate part of nature, as emphasized, for example, by Thure von Uexküll, himself a leading theorist of biosemiotics, in his introductory comment on his father's book on sign theory, comes into play:

> The line drawn between organic and inorganic nature is not determined on the basis of random distinctive features, such as chemical makeup, size, complexity, or the form of the structure in question, but on the basis of a characteristic quality which can first be observed among living things and which is inherent even in the simplest forms of life, the protozoans. This inherent characteristic is the ability of an organism to react to stimuli, not just in a casual-mechanical way, but with its own specific reaction. (von Uexküll, 1992/1987, p. 284)

Whether such a specific reaction and, most importantly, the ownership of the reaction by the organism acting as an individual, as assumed by this distinction, marks a clear boundary between the animate and inanimate parts of nature, can be debated. Nevertheless, this reactivity clearly constitutes the foundation for the development of more complex ways to process environmental signals than those exhibited by protozoans, and ultimately psychological functions, consciousness, and the human mind (see Chap. 6). The neuron developed as specialized cell type, which, among others, serves this function. Yet, this distinct quality in the reaction of forms of life to their environments is what information theory is not able to account for and what alternative approaches are incorporated in their models from a *biological* perspective, in contrast to a biophysical perspective. The molecular level, however, is characterized by biophysical and chemical processes, which, when taken on their own, lack this specific quality of *bios*. Only in the context of the organism-environment system, and specifically the living cell, do biophysical and chemical processes gain significance as contributing factors to signal transmission in the biological sense.

While failing to capture the specificity of biological systems, information theory works well for the identification of signals within noise and is widely used for pattern recognition within a fuzzy data stream. This allows the introduction of preconditions for patterns of interest that are not accounted for in quantitative threshold or correlational models. Its application to neurophysiological measurements in neuroscience, such as EEG, MRI, or PET data, as well as its use in the analysis of genetic and epigenetic data, exactly follows this purpose: reducing the complexity to recognize patterns within the data stream based on some structural preconditions—the information to decipher the code. Whether these recognized patterns are functionally relevant at higher levels of the biological system, where signals are interpreted by the individual in accordance with its interactions with the environment and, thus, the model at the molecular level fits the biological processes, needs to be assessed separately.

The issue of how well models based on information theory criteria actually represent biological processes has been intensely discussed in computational biology (Chanda et al., 2020). Here, the application of information theory in the analysis of transcriptomic or proteomic data is successful in capturing complex data structures that are not accessible using conventional parametric and non-parametric modeling approaches. However, the resulting models are still only approximations of the underlying biophysical and chemical processes, only specifying the likelihood of a certain event or state of a system. The ability of such a model to capture relevant patterns strongly depends on the existing knowledge about the preconditions shaping the information structure, including all involved factors, their possible states, and how they affect the process of information transmission. The degree to which this knowledge is available varies significantly for different biophysical and molecular processes.

Characterizing Gene Expression with Information Theory: The Genon Model

For genetic information, Jost argues that our current knowledge of gene expression allows for such specifications and that information theory can therefore model genetic activity to a high degree of precision (Jost, 2020). Nevertheless, this implies specific conceptualizations of the individual parts that contribute to the information system of the genome. Most importantly, the use of an information theory approach for the description of genetic data implies a clear decision for a functional definition of the "gene" (Scherrer & Jost, 2007). Scherrer and Jost define the gene as "the coding sequence contained within the mRNA and its counterpart after translation, the nascent polypeptide" (Scherrer & Jost, 2007, p. 3). This is in contrast to definitions that describe the gene primarily either as a hereditary unit (see Portin & Wilkins, 2017) or as a comprehensive part of the DNA sequence (Gerstein et al., 2007). In eukaryotes, the gene, according to the functional definition by Scherrer and Jost, is present only at the mRNA level, with, in most cases, no corresponding uninterrupted sequence at the DNA level. While the DNA sequence level contributes to what Scherer and Jost coin the "genon," the "program," which generates the functional mRNA, the "gene." In addition to different parts of the DNA, which, when transcribed, provide the basic sequence of the later-produced functional mRNA, transcription factors, the type and molecular structure of binding sites for these transcription factors, the 3-D chromatin structure, other proteins and RNA molecules, and cell structures necessary for the transcription process and the transportation of the mRNA to the translation site, all contribute to the genon. Modeling the genon allows the determination of the probability of a specific gene to be expressed in a specific cell in the form of mRNA and its translation, for example, into the functional protein. Based on the genon model, the contribution of each part of the genon can be quantitatively determined. This also includes comparative testing of different genons and their gene probabilities. The goal is to specify different genons according to a specific cell state, which are then interpreted as contextual conditions of gene expression. Thus, the genon specifically models the preconditions for a specific gene to be expressed at a certain point in time within a cell. In intermediate periods, when not all preconditions are met, the model allows for the calculation of likelihoods for gene expression based on the maintained parts of the genon, starting from the specific genetic polymorphism to the availability of transcription factors. The genon therefore provides current information about the specific cellular context of the gene and gene expression as an aggregated yet multifactorial measurement.

Jost and Scherrer excluded miRNA interference from their genon model at the time but argued that these could be added when sufficient knowledge about their types, distribution, and function is available (Jost & Scherrer, 2014). Chromatin-bound epigenetic mechanisms, such as DNA methylation patterns and histone modifications, can be easily added to the genon as additional information to further characterize the conditions under which the expression of a specific gene occurs. However, this implies a functional interpretation of epigenetic marks in ways that are directly related to gene expression and requires that their specific contribution to the expression of a particular gene within a specific cell at a specific time point is known. The measurements also require single-cell epigenetic data or a high degree of consistency of this specific epigenetic marker across the analyzed cells or cellular mixtures at the bulk tissue level.

Energy Landscapes and Sensitivity Scores for DNA Methylation Patterns

In a first attempt to apply information theory criteria specifically to epigenetic data, Jenkinson et al. analyzed whole-genome DNA methylation patterns across different cell types, including cancer cells (Jenkinson et al., 2017, 2018, 2019; Koldobskiy et al., 2021). The main purpose of this study was to extract features from the DNA methylation patterns that capture specific functional differences between cell types or differences in the related cellular phenotypes, including different cancer types. For this, they defined the potential energy landscapes of genomic regions based on the degree of DNA methylation stochasticity and distribution across the CpG sites within these regions for each group of cells. Furthermore, they characterized entropy and entropy sensitivity measures for the likelihood of stochastic and environmentally induced methylation changes. Based on these features, they proposed a distance measure to calculate the differences in DNA methylation patterns between cell types, age groups, and disease types in these regions. They showed that the additional informational specificity captured by these measurements allowed for higher discriminative power in cell type identification compared to conventional differential methylation analysis using mean methylation levels or methylation variance across genomic regions (Jenkinson et al., 2017, 2018). They also successfully identified functionally relevant genetic and epigenetic drivers of acute pediatric leukemia (Koldobskiy et al., 2021).

The advantage of this information theory approach compared to conventional DNA methylation measures is that it accounts for the non-randomness of the distribution and methylation levels of CpG sites within the genome. Methylation and demethylation mechanisms likely target groups of CpG sites together because of closeness, functional relatedness, or similarities in chromatin structure. This approach specifically accounts for the CpG architecture of the genome, quantified by CpG density and distance at a specific region of interest, and the biochemical environment related to the methylation machinery (Jenkinson et al., 2017, 2018, 2019). Specifically, Jenkinson et al. accounted for the intercorrelation of DNA methylation of neighboring CpG sites, different CpG densities within genomic regions, and variability in DNA methylation probabilities, indicating the likelihood of methylation changes within cell types or tissue samples. For example, they showed that promoter regions and transcription start sites had higher entropic sensitivity values than exons, introns, and intragenic regions (Jenkinson et al., 2017). This further supports the notion that DNA methylation patterns characterized by these information theory criteria are biologically relevant. In addition, Jenkinson et al. showed for several cell types, including embryonic stem cells, brain cells, CD4+ lymphocytes, colon cells, lung cells, liver cells, and cancer cell types of these tissues, that at genomic sites, which encode developmentally critical genes, the probability distributions of methylation levels and the shape of these distributions are related to functional relevant differences, such as pluripotency in stem cells and fate lineage determination in cancer cells (Jenkinson et al., 2017). Furthermore, the information theory approach revealed that the cell type-specific maintenance and transmission of DNA methylation patterns differ between functionally relevant genomic regions and other parts of the genome. Jenkinson et al. interpreted their findings in the way that DNA methylation in critical regions is facilitated by high-capacity methylation channels providing low methylation stochasticity at high energetic costs, and DNA methylation in less relevant regions is facilitated by low-capacity methylation channels, resulting in higher methylation stochasticity and lower energetic costs (Jenkinson et al., 2017). In addition, environmental exposure influences these measures of epigenetic stochasticity differently along the genome and depending on the cell type. Jenkinson et al. (2017) report that brain cells at the bulk-tissue level showed more sites with entropic sensitivity than other somatic cell types. They concluded that this is an indicator of the neuron's ability to respond to environmental stimuli, such as synaptic activity, via active demethylation, thereby adapting their methylation patterns according to functional affordances.

When evaluating these findings, it is important to note that their measure of methylation stochasticity depends on the probability distribution of DNA methylation patterns within a group of cells from a tissue sample or a specific cell type. Their epigenetic measures do not represent the single-cell epigenome but provide bulk-tissue measurements. Based on their findings, they proposed that functionally relevant properties of the identified DNA methylation patterns may result from groups of neighboring cells and not at the single-cell level:

> The stochastic nature and properties of DNA methylation and their close relationship with chromatin structure raise the possibility that epigenetic information is carried by a population of cells as a whole and that this information helps not only to achieve and maintain a differentiated state but also to mediate developmental plasticity throughout the life of an organism. (Jenkinson et al., 2017, p. 728)

This, however, marks a difference from the information theory approach to gene expression proposed by Scherrer and Jost (2007), which in principle could also be extended to bulk tissue and cell type groups but conceptually assumes the single-cell level.

Overall, the information theory approach of Jenkinson et al. allows the computation of energy landscapes and their potential association with specific methylation patterns at individual genomic sites, joint probabilities of methylation patterns across genomic sites, and marginal probabilities of DNA methylation at individual CpG sites, among other measures. It enhances DNA methylation analysis by accounting for the underlying CpG architecture of the genome, quantified by CpG densities and distances within a genomic unit, and the local biochemical environment provided by the methylation machinery. This provides a more detailed characterization of a specific type of epigenetic information, DNA methylation patterns measured in groups of cells, compared with conventional measures. Thus, it captures further parts of the multilayer structure of epigenetic information and can contribute to further characterizing the context of the neuron and its function at the molecular level.

The Neuron Under the Epigenetic Lens

As we have seen, placing the single neuron under the epigenetic lens results in a blurry picture of a highly individualized epigenetic signature within which epigenetic changes due to developmental, synaptic, and structural-genomic epigenetic mechanisms combined with stochastic changes overlap. Determining functionally relevant structures within epigenetic information at this single neuron level will likely result in an overrepresentation of mechanisms relevant to maintaining cell metabolism and basic cellular function. Epigenetic modifications related to neural function, such as synaptic plasticity, recent activation history, or functional specialization, are difficult to detect against this noise of the regular cell machinery. Regarding the potential indeterminacy and pluripotency at the single neuron level, focusing on the single neuron epigenome is not as informative as suggested by our traditional view of the neuron as a basic unit and the ultimate source of neural function at higher network levels. This is in line with the equifinality and stochasticity of biological processes. We cannot expect functional changes to be reliably registered at every exposed neuron. Cells, and even neurons, do not work with the necessary rigor to ensure this form of reliability. Systems biology has understood and modeled this issue quite well for quite some time (Wagner, 1999; Wright, 1979). Nonetheless, this notion still poses, as it seems, a major challenge to our concepts, models, and theories in psychology. To endure this uncertainty regarding our understanding of the material foundations of our mind goes straight against our cultural heritage in which the atomistic version of the mechanistic worldview dominates. Clearly, every attempt to pinpoint down all molecular interactions within a specific single neuron will further advance our understanding of these cells and their cellular function and survival. It will just not contribute much to the determination of the neural basis of a specific psychological function.

The image clearly sharpens when we zoom out slightly and focus on the level of cell ensembles, small neural populations, cell groups, or local sites of bulk tissue. Similar to the fact that a single skin cell is not vital to the protection that the skin provides to our body, while larger wounds clearly pose health threats from small local infections up to lasting injuries, epigenetic signatures that are established and maintained across larger groups of neuronal cells or brain tissue will have a greater likelihood of being functionally relevant. Nevertheless, at this level, scores, gradients, and typical patterns must be identified to differentiate the layers aggregated in this epigenetic signature. As we have seen mainly for DNA methylation, the already available knowledge about the information structure of epigenetic data, the inherent patterns, can be used to trace differences and changes at a multicellular or tissue-bulk level and in relation to the underlying genomic structure in a way that is informative for the detection of potential functional differences. However, we need to keep in mind that this does not mean that these patterns are more than discriminative biomarkers. Whether and how they are mechanistically related to these functional differences needs to be assessed in future studies.

However, cell- or tissue-specific epigenetic profiles related to cellular homeostasis processes provide a powerful information filter that accounts for the typical background epigenetic state. Additionally, epigenetic clocks have already established rough measures of steady stochastic changes related to chronological aging. Moreover, the information theory approach allows the identification of additional layers of epigenetic information in a specific tissue, for example, by identifying genomic loci with higher and lower entropic sensitivity of DNA methylation in general and related to specific forms of exposure. Together, these provide a powerful set of information-decoding devices for the interpretation of epigenetic data. They can be used to identify changes due to development and cellular differentiation, to characterize critical and sensible periods, for example, according to changes in the energy landscape or entropic sensitivity, and to detect stochastic changes induced by diseases or functional failure, such as lesions, strokes, hormonal influences, and exposure to toxins.

What does this add to the picture of neural activity patterns that we cannot already detect with single- or multicell recordings or at the level of small neural populations? Although all measurements of the epigenetic signature are always snapshots in time, capturing only one moment in a highly dynamic process, we can assume that parts of this snapshot are relatively stable, perhaps even across the lifespan. This is the main difference between these molecular epigenetic markers and measurements of neural activity or gene expression. The picture we see through the molecular epigenetic lens is not entirely in constant flux and does not change in total with every synaptic or other cellular input. This would not be effective in terms of the energetic costs of epigenetic modifications. As the exemplary analysis of the information structure of DNA methylation states by Jenkinson et al. (2017, 2018, 2019) showed, we can even use estimates of these energy landscapes to calculate the stability or likelihood of changes in DNA methylation patterns at specific genomic sites at a multicellular level. One promising approach is to identify genomic sites with either relatively stable or highly dynamic epigenetic signatures in the context of a specific functional failure or environmental exposure, as well as to calculate stability ranges related to exposure. Here, the energetic costs of stabilization versus change could function as possible indicators of exposure effects.

Overall, we need to assume that genetic and epigenetic activity protects neurons from the constant influx of information coming from their immediate environment, with threshold models also being likely for chromatin remodeling processes and RNA interference. Thus, referring to Waddington's picture of the epigenetic landscape, for an input to result in enduring changes, the system must be pushed strong enough to land in a different pathway equilibrium. Nonetheless, the identification of epigenetic modifications, either because of exposure to a specific environmental signal, a specific tendency to modify in a certain developmental period, or both, does not automatically imply the functional relevance of these changes at the level of neural function, either at the level of small neural networks or at the single neuron level. However, the identification of specific forms and sites of modification will help direct research efforts. Their most valuable addition will be to provide information about the developmental, activation, and exposure histories of neural tissue or small brain nuclei, which may allow us to further understand the characteristics and potential constraints of neural activity patterns resulting from a specific developmental pathway.

Thus, although the single neuron still escapes the epigenetic lens, neuroepigenetic measures allow us to zoom at a potentially functionally relevant level. They provide traces of the local developmental and functional history of neural activity as preserved in the local brain tissue in the form of this specific molecular layer, contributing to the characterization of the context of the single neuron. Due to their relative stability, they have the potential to provide us with a biomarker of past exposure and activation that is not detectable with momentary measurements at the neural activity level. However, to harvest and decipher this potentially rich layer of epigenetic information, we need to further characterize these epigenetic signatures and develop markers and filters that help us best calibrate our molecular epigenetic lens.

References

Agboola, O. S., Hu, X., Shan, Z., Wu, Y., & Lei, L. (2021). Brain organoid: A 3D technology for investigating cellular composition and interactions in human neurological development and disease models in vitro. *Stem Cell Research & Therapy, 12*(1), 430. https://doi.org/10.1186/s13287-021-02369-8

Amiri, A., Coppola, G., Scuderi, S., Wu, F., Roychowdhury, T., Liu, F., Pochareddy, S., Shin, Y., Safi, A., Song, L., Zhu, Y., Sousa, A. M. M., Gerstein, M., Crawford, G. E., Sestan, N., Abyzov, A., Vaccarino, F. M., & PsychENCODE Consortium. (2018). Transcriptome and epigenome landscape of human cortical development modeled in organoids. *Science (New York, N.Y.), 362*(6420), eaat6720. https://doi.org/10.1126/science.aat6720

Azzi, A., Dallmann, R., Casserly, A., Rehrauer, H., Patrignani, A., Maier, B., Kramer, A., & Brown, S. A. (2014). Circadian behavior is light-reprogrammed by plastic DNA methylation. *Nature Neuroscience, 17*(3), 377–382. https://doi.org/10.1038/nn.3651

Backes, E., & Hemby, S. E. (2003). Discrete cell gene profiling of ventral tegmental dopamine neurons after acute and chronic cocaine self-administration. *The Journal of Pharmacology and Experimental Therapeutics, 307*(2), 450–459. https://doi.org/10.1124/jpet.103.054965

Bearer, E. L., & Mulligan, B. S. (2018). Epigenetic changes associated with early life experiences: Saliva, a biospecimen for DNA methylation signatures. *Current Genomics, 19*(8), 676–698. https://doi.org/10.2174/1389202919666180307150508

Bell, C. G., Lowe, R., Adams, P. D., Baccarelli, A. A., Beck, S., Bell, J. T., Christensen, B. C., Gladyshev, V. N., Heijmans, B. T., Horvath, S., Ideker, T., Issa, J.-P. J., Kelsey, K. T., Marioni, R. E., Reik, W., Relton, C. L., Schalkwyk, L. C., Teschendorff, A. E., Wagner, W., et al. (2019). DNA methylation aging clocks: Challenges and recommendations. *Genome Biology, 20*(1), 249. https://doi.org/10.1186/s13059-019-1824-y

Bergsma, T., & Rogaeva, E. (2020). DNA methylation clocks and their predictive capacity for aging phenotypes and healthspan. *Neuroscience Insights, 15*, 2633105520942221. https://doi.org/10.1177/2633105520942221

Bheda, P., & Schneider, R. (2014). Epigenetics reloaded: The single-cell revolution. *Trends in Cell Biology, 24*(11), 712–723. https://doi.org/10.1016/j.tcb.2014.08.010

Bibikova, M., Barnes, B., Tsan, C., Ho, V., Klotzle, B., Le, J. M., Delano, D., Zhang, L., Schroth, G. P., Gunderson, K. L., Fan, J.-B., & Shen, R. (2011). High density DNA methylation array with single CpG site resolution. *Genomics, 98*(4), 288–295. https://doi.org/10.1016/j.ygeno.2011.07.007

Bock, J., Wainstock, T., Braun, K., & Segal, M. (2015). Stress in utero: Prenatal programming of brain plasticity and cognition. *Biological Psychiatry, 78*(5), 315–326. https://doi.org/10.1016/j.biopsych.2015.02.036

Bocklandt, S., Lin, W., Sehl, M. E., Sánchez, F. J., Sinsheimer, J. S., Horvath, S., & Vilain, E. (2011). Epigenetic predictor of age. *PLoS One, 6*(6), e14821. https://doi.org/10.1371/journal.pone.0014821

Borst, A., & Theunissen, F. E. (1999). Information theory and neural coding. *Nature Neuroscience, 2*(11), Article 11. https://doi.org/10.1038/14731

Brägelmann, J., & Lorenzo Bermejo, J. (2018). A comparative analysis of cell-type adjustment methods for epigenome-wide association studies based on simulated and real data sets. *Briefings in Bioinformatics, 20*(6), 2055–2065. https://doi.org/10.1093/bib/bby068

Brette, R. (2019). Is coding a relevant metaphor for the brain? *Behavioral and Brain Sciences, 42*, e215. https://doi.org/10.1017/S0140525X19000049

Bühler, K. (1960). Das Gestaltprinzip im Leben des Menschen und der Tiere. https://ids-pub.bsz-bw.de/frontdoor/index/index/docId/5913

Bühler, K. (1999). *Sprachtheorie: Die Darstellungsfunktion der Sprache* (3. Aufl., ungekürzter Neudr. d. Ausg. Jena, Fischer, 1934). Lucius und Lucius.

Chanda, P., Costa, E., Hu, J., Sukumar, S., Van Hemert, J., & Walia, R. (2020). Information theory in computational biology: Where we stand today. *Entropy (Basel, Switzerland), 22*(6), 627. https://doi.org/10.3390/e22060627

Clark, S. J., Lee, H. J., Smallwood, S. A., Kelsey, G., & Reik, W. (2016). Single-cell epigenomics: Powerful new methods for understanding gene regulation and cell identity. *Genome Biology, 17*(1), 72. https://doi.org/10.1186/s13059-016-0944-x

Cronbach, J. L. (1955). On the non-rational application of information measures in psychology. In H. Quastler (Ed.), *Information theory in psychology* (pp. 14–26). The Free Press.

de Lima Camillo, L. P., Lapierre, L. R., & Singh, R. (2022). A pan-tissue DNA-methylation epigenetic clock based on deep learning. *Npj Aging, 8*(1), Article 1. https://doi.org/10.1038/s41514-022-00085-y

de Ruyter van Steveninck, R., Bialek, W., & Barlow, H. B. (1997). Real-time performance of a movement-sensitive neuron in the blowfly visual system: Coding and information transfer in short spike sequences. *Proceedings of the Royal Society of London. Series B. Biological Sciences, 234*(1277), 379–414. https://doi.org/10.1098/rspb.1988.0055

Di Lullo, E., & Kriegstein, A. R. (2017). The use of brain organoids to investigate neural development and disease. *Nature Reviews Neuroscience, 18*(10), Article 10. https://doi.org/10.1038/nrn.2017.107

Dieckmann, L., Cruceanu, C., Lahti-Pulkkinen, M., Lahti, J., Kvist, T., Laivuori, H., Sammallahti, S., Villa, P. M., Suomalainen-König, S., Rancourt, R. C., Plagemann, A., Henrich, W., Eriksson, J. G., Kajantie, E., Entringer, S., Braun, T., Räikkönen, K., Binder, E. B., & Czamara, D. (2022). Reliability of a novel approach for reference-based cell type estimation in human placental DNA methylation studies. *Cellular and Molecular Life Sciences, 79*(2), 115. https://doi.org/10.1007/s00018-021-04091-3

Dimitrov, A. G., Lazar, A. A., & Victor, J. D. (2011). Information theory in neuroscience. *Journal of Computational Neuroscience, 30*(1), 1–5. https://doi.org/10.1007/s10827-011-0314-3

Fraga, M. F., Ballestar, E., Paz, M. F., Ropero, S., Setien, F., Ballestar, M. L., Heine-Suñer, D., Cigudosa, J. C., Urioste, M., Benitez, J., Boix-Chornet, M., Sanchez-Aguilera, A., Ling, C., Carlsson, E., Poulsen, P., Vaag, A., Stephan, Z., Spector, T. D., Wu, Y.-Z., et al. (2005). Epigenetic differences arise during the lifetime of monozygotic twins. *Proceedings of the National Academy of Sciences of the United States of America, 102*(30), 10604. https://doi.org/10.1073/pnas.0500398102

Fransquet, P. D., Wrigglesworth, J., Woods, R. L., Ernst, M. E., & Ryan, J. (2019). The epigenetic clock as a predictor of disease and mortality risk: A systematic review and meta-analysis. *Clinical Epigenetics, 11*(1), 62. https://doi.org/10.1186/s13148-019-0656-7

Galkin, F., Mamoshina, P., Kochetov, K., Sidorenko, D., & Zhavoronkov, A. (2021). DeepMAge: A methylation aging clock developed with deep learning. *Aging and Disease, 12*(5), 1252–1262. https://doi.org/10.14336/AD.2020.1202

Gerstein, M. B., Bruce, C., Rozowsky, J. S., Zheng, D., Du, J., Korbel, J. O., Emanuelsson, O., Zhang, Z. D., Weissman, S., & Snyder, M. (2007). What is a gene, post-ENCODE? History and updated definition. *Genome Research, 17*(6), 669–681. https://doi.org/10.1101/gr.6339607

Gorkin, D. U., Qiu, Y., Hu, M., Fletez-Brant, K., Liu, T., Schmitt, A. D., Noor, A., Chiou, J., Gaulton, K. J., Sebat, J., Li, Y., Hansen, K. D., & Ren, B. (2019). Common DNA sequence variation influences 3-dimensional conformation of the human genome. *Genome Biology, 20*(1), 255. https://doi.org/10.1186/s13059-019-1855-4

Grodstein, F., Lemos, B., Yu, L., Iatrou, A., De Jager, P. L., & Bennett, D. A. (2021). Characteristics of epigenetic clocks across blood and brain tissue in older women and men. *Frontiers in Neuroscience, 14*. https://www.frontiersin.org/articles/10.3389/fnins.2020.555307

Hannum, G., Guinney, J., Zhao, L., Zhang, L., Hughes, G., Sadda, S., Klotzle, B., Bibikova, M., Fan, J.-B., Gao, Y., Deconde, R., Chen, M., Rajapakse, I., Friend, S., Ideker, T., & Zhang, K. (2013). Genome-wide methylation profiles reveal quantitative views of human aging rates. *Molecular Cell, 49*(2), 359–367. https://doi.org/10.1016/j.molcel.2012.10.016

Hayashi-Takanaka, Y., Kina, Y., Nakamura, F., Becking, L. E., Nakao, Y., Nagase, T., Nozaki, N., & Kimura, H. (2020). Histone modification dynamics as revealed by multicolor immunofluorescence-based single-cell analysis. *Journal of Cell Science, 133*(14), jcs243444. https://doi.org/10.1242/jcs.243444

Heiss, J. A., Brennan, K. J., Baccarelli, A. A., Téllez-Rojo, M. M., Estrada-Gutiérrez, G., Wright, R. O., & Just, A. C. (2019). Battle of epigenetic proportions: Comparing Illumina's EPIC methylation microarrays and TruSeq targeted bisulfite sequencing. *Epigenetics, 15*(1–2), 174–182. https://doi.org/10.1080/15592294.2019.1656159

Hempel, C. M., Sugino, K., & Nelson, S. B. (2007). A manual method for the purification of fluorescently labeled neurons from the mammalian brain. *Nature Protocols, 2*(11), 2924–2929. https://doi.org/10.1038/nprot.2007.416

Hoffmeyer, J. (2008). *Biosemiotics: An examination into the signs of life and the life of signs.* University of Scranton Press.

Horvath, S. (2013). DNA methylation age of human tissues and cell types. *Genome Biology, 14*(10), R115. https://doi.org/10.1186/gb-2013-14-10-r115

Huang, Y., Yan, J., Hou, J., Fu, X., Li, L., & Hou, Y. (2015). Developing a DNA methylation assay for human age prediction in blood and bloodstain. *Forensic Science International. Genetics, 17*, 129–136. https://doi.org/10.1016/j.fsigen.2015.05.007

Huang, R.-C., Lillycrop, K. A., Beilin, L. J., Godfrey, K. M., Anderson, D., Mori, T. A., Rauschert, S., Craig, J. M., Oddy, W. H., Ayonrinde, O. T., Pennell, C. E., Holbrook, J. D., & Melton, P. E. (2019). Epigenetic age acceleration in adolescence associates with BMI, inflammation, and risk score for middle age cardiovascular disease. *The Journal of Clinical Endocrinology and Metabolism, 104*(7), 3012–3024. https://doi.org/10.1210/jc.2018-02076

Jaffe, A. E., & Irizarry, R. A. (2014). Accounting for cellular heterogeneity is critical in epigenome-wide association studies. *Genome Biology, 15*(2), R31. https://doi.org/10.1186/gb-2014-15-2-r31

Jang, H. S., Shin, W. J., Lee, J. E., & Do, J. T. (2017). CpG and non-CpG methylation in epigenetic gene regulation and brain function. *Genes, 8*(6), 148. https://doi.org/10.3390/genes8060148

Jenkinson, G., Pujadas, E., Goutsias, J., & Feinberg, A. P. (2017). Potential energy landscapes identify the information-theoretic nature of the epigenome. *Nature Genetics, 49*(5), Article 5. https://doi.org/10.1038/ng.3811

Jenkinson, G., Abante, J., Feinberg, A. P., & Goutsias, J. (2018). An information-theoretic approach to the modeling and analysis of whole-genome bisulfite sequencing data. *BMC Bioinformatics, 19*(1), 87. https://doi.org/10.1186/s12859-018-2086-5

Jenkinson, G., Abante, J., Koldobskiy, M. A., Feinberg, A. P., & Goutsias, J. (2019). Ranking genomic features using an information-theoretic measure of epigenetic discordance. *BMC Bioinformatics, 20*(1), 175. https://doi.org/10.1186/s12859-019-2777-6

Johnstone, S. E., Gladyshev, V. N., Aryee, M. J., & Bernstein, B. E. (2022). Epigenetic clocks, aging, and cancer. *Science, 378*(6626), 1276–1277. https://doi.org/10.1126/science.abn4009

Jones, P. A., & Baylin, S. B. (2007). The epigenomics of cancer. *Cell, 128*(4), 683–692. https://doi.org/10.1016/j.cell.2007.01.029

Jost, J. (2020). Biological information. *Theory in Biosciences = Theorie in Den Biowissenschaften, 139*(4), 361–370. https://doi.org/10.1007/s12064-020-00327-1

Jost, J. (2021). Information theory and consciousness. *Frontiers in Applied Mathematics and Statistics, 7*. https://www.frontiersin.org/articles/10.3389/fams.2021.641239

Jost, J., & Scherrer, K. (2014). Information theory, gene expression, and combinatorial regulation: A quantitative analysis. *Theory in Biosciences = Theorie in Den Biowissenschaften, 133*(1), 1–21. https://doi.org/10.1007/s12064-013-0182-7

Kadoshima, T., Sakaguchi, H., Nakano, T., Soen, M., Ando, S., Eiraku, M., & Sasai, Y. (2013). Self-organization of axial polarity, inside-out layer pattern, and species-specific progenitor dynamics in human ES cell-derived neocortex. *Proceedings of the National Academy of Sciences of the United States of America, 110*(50), 20284–20289. https://doi.org/10.1073/pnas.1315710110

Karemaker, I. D., & Vermeulen, M. (2018). Single-cell DNA methylation profiling: Technologies and biological applications. *Trends in Biotechnology, 36*(9), 952–965. https://doi.org/10.1016/j.tibtech.2018.04.002

Khodadadi, E., Fahmideh, L., Khodadadi, E., Dao, S., Yousefi, M., Taghizadeh, S., Asgharzadeh, M., Yousefi, B., & Kafil, H. S. (2021). Current advances in DNA methylation analysis methods. *BioMed Research International, 2021*, e8827516. https://doi.org/10.1155/2021/8827516

Knight, A. K., Craig, J. M., Theda, C., Bækvad-Hansen, M., Bybjerg-Grauholm, J., Hansen, C. S., Hollegaard, M. V., Hougaard, D. M., Mortensen, P. B., Weinsheimer, S. M., Werge, T. M., Brennan, P. A., Cubells, J. F., Newport, D. J., Stowe, Z. N., Cheong, J. L. Y., Dalach, P., Doyle, L. W., Loke, Y. J., et al. (2016). An epigenetic clock for gestational age at birth based on blood methylation data. *Genome Biology, 17*(1), 206. https://doi.org/10.1186/s13059-016-1068-z

Koch, C. (1996). A neuronal correlate of consciousness? *Current Biology, 6*(5), 492. https://doi.org/10.1016/S0960-9822(02)00519-5

Koch, C. (2019). *The feeling of life itself: Why consciousness is widespread but can't be computed.* MIT Press.

Koch, C. M., & Wagner, W. (2011). Epigenetic-aging-signature to determine age in different tissues. *Aging, 3*(10), 1018–1027. https://doi.org/10.18632/aging.100395

Koldobskiy, M. A., Jenkinson, G., Abante, J., DiBlasi, V. A. R., Zhou, W., Pujadas, E., Idrizi, A., Tryggvadottir, R., Callahan, C., Bonifant, C. L., Rabin, K. R., Brown, P. A., Ji, H., Goutsias, J., & Feinberg, A. P. (2021). An information-theory analysis of DNA methylation identifies converging genetic and epigenetic drivers of paediatric acute lymphoblastic leukaemia. *Nature Biomedical Engineering, 5*(4), 360–376. https://doi.org/10.1038/s41551-021-00703-2

Kumsta, R. (2019). The role of epigenetics for understanding mental health difficulties and its implications for psychotherapy research. *Psychology and Psychotherapy: Theory, Research and Practice, 92*, 190–207. https://doi.org/10.1111/papt.12227

Laming, D. (2001). Statistical information, uncertainty, and Bayes' Theorem: Some applications in experimental psychology. In S. Benferhat & P. Besnard (Eds.), *Symbolic and quantitative approaches to reasoning with uncertainty* (pp. 635–646). Springer. https://doi.org/10.1007/3-540-44652-4_56

Lancaster, M. A., Renner, M., Martin, C.-A., Wenzel, D., Bicknell, L. S., Hurles, M. E., Homfray, T., Penninger, J. M., Jackson, A. P., & Knoblich, J. A. (2013). Cerebral organoids model human brain development and microcephaly. *Nature, 501*(7467), 373–379. https://doi.org/10.1038/nature12517

Lappalainen, T., & Greally, J. M. (2017). Associating cellular epigenetic models with human phenotypes. *Nature Reviews Genetics, 18*(7), Article 7. https://doi.org/10.1038/nrg.2017.32

Laszlo, A. H., Derrington, I. M., Brinkerhoff, H., Langford, K. W., Nova, I. C., Samson, J. M., Bartlett, J. J., Pavlenok, M., & Gundlach, J. H. (2013). Detection and mapping of 5-methylcytosine and 5-hydroxymethylcytosine with nanopore MspA. *Proceedings of the National Academy of Sciences, 110*(47), 18904–18909. https://doi.org/10.1073/pnas.1310240110

Laurent, L., Wong, E., Li, G., Huynh, T., Tsirigos, A., Ong, C. T., Low, H. M., Kin Sung, K. W., Rigoutsos, I., Loring, J., & Wei, C.-L. (2010). Dynamic changes in the human methylome during differentiation. *Genome Research, 20*(3), 320–331. https://doi.org/10.1101/gr.101907.109

Law, P.-P., & Holland, M. L. (2019). DNA methylation at the crossroads of gene and environment interactions. *Essays in Biochemistry, 63*(6), 717–726. https://doi.org/10.1042/EBC20190031

Levine, M. E., Lu, A. T., Quach, A., Chen, B. H., Assimes, T. L., Bandinelli, S., Hou, L., Baccarelli, A. A., Stewart, J. D., Li, Y., Whitsel, E. A., Wilson, J. G., Reiner, A. P., Aviv, A., Lohman, K., Liu, Y., Ferrucci, L., & Horvath, S. (2018). An epigenetic biomarker of aging for lifespan and healthspan. *Aging (Albany NY), 10*(4), 573–591. https://doi.org/10.18632/aging.101414

Li, A., Koch, Z., & Ideker, T. (2022a). Epigenetic aging: Biological age prediction and informing a mechanistic theory of aging. *Journal of Internal Medicine, 292*(5), 733–744. https://doi.org/10.1111/joim.13533

Li, A., Mueller, A., English, B., Arena, A., Vera, D., Kane, A. E., & Sinclair, D. A. (2022b). Novel feature selection methods for construction of accurate epigenetic clocks. *PLoS Computational Biology, 18*(8), e1009938. https://doi.org/10.1371/journal.pcbi.1009938

Liang, L., Chang, Y., Lu, J., Wu, X., Liu, Q., Zhang, W., Su, X., & Zhang, B. (2019). Global methylomic and transcriptomic analyses reveal the broad participation of DNA methylation in daily gene expression regulation of Populus trichocarpa. *Frontiers in Plant Science, 10*, 243. https://doi.org/10.3389/fpls.2019.00243

Lister, R., Pelizzola, M., Dowen, R. H., Hawkins, R. D., Hon, G., Tonti-Filippini, J., Nery, J. R., Lee, L., Ye, Z., Ngo, Q.-M., Edsall, L., Antosiewicz-Bourget, J., Stewart, R., Ruotti, V., Millar, A. H., Thomson, J. A., Ren, B., & Ecker, J. R. (2009). Human DNA methylomes at base resolution show widespread epigenomic differences. *Nature, 462*(7271), 315–322. https://doi.org/10.1038/nature08514

Liu, A. (2010). Laser capture microdissection in the tissue biorepository. *Journal of Biomolecular Techniques: JBT, 21*(3), 120–125.

Liu, Y., Aryee, M. J., Padyukov, L., Fallin, M. D., Hesselberg, E., Runarsson, A., Reinius, L., Acevedo, N., Taub, M., Ronninger, M., Shchetynsky, K., Scheynius, A., Kere, J., Alfredsson, L., Klareskog, L., Ekström, T. J., & Feinberg, A. P. (2013). Epigenome-wide association data implicate DNA methylation as an intermediary of genetic risk in Rheumatoid Arthritis. *Nature Biotechnology, 31*(2), 142–147. https://doi.org/10.1038/nbt.2487

Lobo, M. K., Karsten, S. L., Gray, M., Geschwind, D. H., & Yang, X. W. (2006). FACS-array profiling of striatal projection neuron subtypes in juvenile and adult mouse brains. *Nature Neuroscience, 9*(3), 443–452. https://doi.org/10.1038/nn1654

Lowe, R., & Rakyan, V. K. (2014). Correcting for cell-type composition bias in epigenome-wide association studies. *Genome Medicine, 6*(3), 23. https://doi.org/10.1186/gm540

Lu, A. T., Quach, A., Wilson, J. G., Reiner, A. P., Aviv, A., Raj, K., Hou, L., Baccarelli, A. A., Li, Y., Stewart, J. D., Whitsel, E. A., Assimes, T. L., Ferrucci, L., & Horvath, S. (2019). DNA methylation GrimAge strongly predicts life-span and healthspan. *Aging, 11*(2), 303–327. https://doi.org/10.18632/aging.101684

Luce, R. D. (2003). Whatever happened to information theory in psychology? *Review of General Psychology, 7*, 183–188. https://doi.org/10.1037/1089-2680.7.2.183

MacKay, D. M., & McCulloch, W. S. (1952). The limiting information capacity of a neuronal link. *The Bulletin of Mathematical Biophysics, 14*(2), 127–135. https://doi.org/10.1007/BF02477711

Martins, J., Czamara, D., Sauer, S., Rex-Haffner, M., Dittrich, K., Dörr, P., de Punder, K., Overfeld, J., Knop, A., Dammering, F., Entringer, S., Winter, S. M., Buss, C., Heim, C., & Binder, E. B. (2021). Childhood adversity correlates with stable changes in DNA methylation trajectories in children and converges with epigenetic signatures of prenatal stress. *Neurobiology of Stress, 15*, 100336. https://doi.org/10.1016/j.ynstr.2021.100336

McGregor, K., Bernatsky, S., Colmegna, I., Hudson, M., Pastinen, T., Labbe, A., & Greenwood, C. M. T. (2016). An evaluation of methods correcting for cell-type heterogeneity in DNA methylation studies. *Genome Biology, 17*(1), 84. https://doi.org/10.1186/s13059-016-0935-y

Morales Berstein, F., McCartney, D. L., Lu, A. T., Tsilidis, K. K., Bouras, E., Haycock, P., Burrows, K., Phipps, A. I., Buchanan, D. D., Cheng, I., Martin, R. M., Davey Smith, G., Relton, C. L., Horvath, S., Marioni, R. E., Richardson, T. G., Richmond, R. C., & the PRACTICAL consortium. (2022). Assessing the causal role of epigenetic clocks in the development of multiple cancers: A Mendelian randomization study. *eLife, 11*, e75374. https://doi.org/10.7554/eLife.75374

Niu, Y., DesMarais, T. L., Tong, Z., Yao, Y., & Costa, M. (2015). Oxidative stress alters global histone modification and DNA methylation. *Free Radical Biology and Medicine, 82*, 22–28. https://doi.org/10.1016/j.freeradbiomed.2015.01.028

O'Neill, H., Lee, H., Gupta, I., Rodger, E. J., & Chatterjee, A. (2022). Single-cell DNA methylation analysis in cancer. *Cancers, 14*(24), Article 24. https://doi.org/10.3390/cancers14246171

Oblak, L., van der Zaag, J., Higgins-Chen, A. T., Levine, M. E., & Boks, M. P. (2021). A systematic review of biological, social and environmental factors associated with epigenetic clock acceleration. *Ageing Research Reviews, 69*, 101348. https://doi.org/10.1016/j.arr.2021.101348

Oh, E. S., & Petronis, A. (2021). Origins of human disease: The chrono-epigenetic perspective. *Nature Reviews Genetics, 22*(8), Article 8. https://doi.org/10.1038/s41576-021-00348-6

Oh, G., Ebrahimi, S., Carlucci, M., Zhang, A., Nair, A., Groot, D. E., Labrie, V., Jia, P., Oh, E. S., Jeremian, R. H., Susic, M., Shrestha, T. C., Ralph, M. R., Gordevičius, J., Koncevičius, K., & Petronis, A. (2018). Cytosine modifications exhibit circadian oscillations that are involved in epigenetic diversity and aging. *Nature Communications, 9*(1), Article 1. https://doi.org/10.1038/s41467-018-03073-7

Oh, G., Koncevičius, K., Ebrahimi, S., Carlucci, M., Groot, D. E., Nair, A., Zhang, A., Kriščiūnas, A., Oh, E. S., Labrie, V., Wong, A. H. C., Gordevičius, J., Jia, P., Susic, M., & Petronis, A. (2019). Circadian oscillations of cytosine modification in humans contribute to epigenetic variability, aging, and complex disease. *Genome Biology, 20*(1), 2. https://doi.org/10.1186/s13059-018-1608-9

Olde Loohuis, N. F. M., Kos, A., Martens, G. J. M., Van Bokhoven, H., Nadif Kasri, N., & Aschrafi, A. (2012). MicroRNA networks direct neuronal development and plasticity. *Cellular and Molecular Life Sciences, 69*(1), 89–102. https://doi.org/10.1007/s00018-011-0788-1

Önder, Ö., Sidoli, S., Carroll, M., & Garcia, B. A. (2015). Progress in epigenetic histone modification analysis by mass spectrometry for clinical investigations. *Expert Review of Proteomics, 12*(5), 499–517. https://doi.org/10.1586/14789450.2015.1084231

Oyama, S. (2000). *The ontogeny of information: Developmental systems and evolution* (2nd ed., rev.and expanded). Duke University Press.

Paul, D. S., Teschendorff, A. E., Dang, M. A. N., Lowe, R., Hawa, M. I., Ecker, S., Beyan, H., Cunningham, S., Fouts, A. R., Ramelius, A., Burden, F., Farrow, S., Rowlston, S., Rehnstrom, K., Frontini, M., Downes, K., Busche, S., Cheung, W. A., Ge, B., et al. (2016). Increased DNA methylation variability in type 1 diabetes across three immune effector cell types. *Nature Communications, 7*(1), Article 1. https://doi.org/10.1038/ncomms13555

Portin, P., & Wilkins, A. (2017). The evolving definition of the term "Gene". *Genetics, 205*(4), 1353–1364. https://doi.org/10.1534/genetics.116.196956

Qi, L., & Teschendorff, A. E. (2022). Cell-type heterogeneity: Why we should adjust for it in epigenome and biomarker studies. *Clinical Epigenetics, 14*(1), 31. https://doi.org/10.1186/s13148-022-01253-3

Qian, X., Nguyen, H. N., Song, M. M., Hadiono, C., Ogden, S. C., Hammack, C., Yao, B., Hamersky, G. R., Jacob, F., Zhong, C., Yoon, K.-J., Jeang, W., Lin, L., Li, Y., Thakor, J., Berg, D. A., Zhang, C., Kang, E., Chickering, M., et al. (2016). Brain-region-specific organoids using mini-bioreactors for modeling ZIKV exposure. *Cell, 165*(5), 1238–1254. https://doi.org/10.1016/j.cell.2016.04.032

Rahmani, E., Schweiger, R., Rhead, B., Criswell, L. A., Barcellos, L. F., Eskin, E., Rosset, S., Sankararaman, S., & Halperin, E. (2019). Cell-type-specific resolution epigenetics without the need for cell sorting or single-cell biology. *Nature Communications, 10*(1), Article 1. https://doi.org/10.1038/s41467-019-11052-9

Rand, A. C., Jain, M., Eizenga, J. M., Musselman-Brown, A., Olsen, H. E., Akeson, M., & Paten, B. (2017). Mapping DNA methylation with high-throughput nanopore sequencing. *Nature Methods, 14*(4), Article 4. https://doi.org/10.1038/nmeth.4189

Reinius, L. E., Acevedo, N., Joerink, M., Pershagen, G., Dahlén, S.-E., Greco, D., Söderhäll, C., Scheynius, A., & Kere, J. (2012). Differential DNA methylation in purified human blood cells: Implications for cell lineage and studies on disease susceptibility. *PLoS One, 7*(7), e41361. https://doi.org/10.1371/journal.pone.0041361

Riffo-Campos, Á. L., Castillo, J., Tur, G., González-Figueroa, P., Georgieva, E. I., Rodríguez, J. L., López-Rodas, G., Rodrigo, M. I., & Franco, L. (2015). Nucleosome-specific, time-dependent changes in histone modifications during activation of the early growth response 1 (Egr1) gene. *The Journal of Biological Chemistry, 290*(1), 197–208. https://doi.org/10.1074/jbc.M114.579292

Ruberti, F., Barbato, C., & Cogoni, C. (2012). Targeting microRNAs in neurons: Tools and perspectives. *Experimental Neurology, 235*(2), 419–426. https://doi.org/10.1016/j.expneurol.2011.10.031

Rulands, S., Lee, H. J., Clark, S. J., Angermueller, C., Smallwood, S. A., Krueger, F., Mohammed, H., Dean, W., Nichols, J., Rugg-Gunn, P., Kelsey, G., Stegle, O., Simons, B. D., & Reik, W. (2018). Genome-scale oscillations in DNA methylation during exit from pluripotency. *Cell Systems, 7*(1), 63–76.e12. https://doi.org/10.1016/j.cels.2018.06.012

Sayood, K. (2018). Information theory and cognition: A review. *Entropy, 20*(9), 706. https://doi.org/10.3390/e20090706

Scherrer, K. (2018). Primary transcripts: From the discovery of RNA processing to current concepts of gene expression – Review. *Experimental Cell Research, 373*(1–2), 1–33. https://doi.org/10.1016/j.yexcr.2018.09.011

Scherrer, K., & Jost, J. (2007). The gene and the genon concept: A functional and information-theoretic analysis. *Molecular Systems Biology, 3*, 87. https://doi.org/10.1038/msb4100123

Schöner, G., Reimann, H., & Lins, J. (2015). Neural dynamics. In G. Schöner, J. Spencer, & D. Research Group (Eds.), *Dynamic thinking: A primer on dynamic field theory* (p. 0). Oxford University Press. https://doi.org/10.1093/acprof:oso/9780199300563.003.0001

Shannon, C. E. (1948). A mathematical theory of communication. *The Bell System Technical Journal, 27*(3), 379–423. https://doi.org/10.1002/j.1538-7305.1948.tb01338.x

Shireby, G. L., Davies, J. P., Francis, P. T., Burrage, J., Walker, E. M., Neilson, G. W. A., Dahir, A., Thomas, A. J., Love, S., Smith, R. G., Lunnon, K., Kumari, M., Schalkwyk, L. C., Morgan, K., Brookes, K., Hannon, E., & Mill, J. (2020). Recalibrating the epigenetic clock: Implications for assessing biological age in the human cortex. *Brain, 143*(12), 3763–3775. https://doi.org/10.1093/brain/awaa334

Simpkin, A. J., Hemani, G., Suderman, M., Gaunt, T. R., Lyttleton, O., Mcardle, W. L., Ring, S. M., Sharp, G. C., Tilling, K., Horvath, S., Kunze, S., Peters, A., Waldenberger, M., Ward-Caviness, C., Nohr, E. A., Sørensen, T. I. A., Relton, C. L., & Smith, G. D. (2016). Prenatal and early life influences on epigenetic age in children: A study of mother-offspring pairs from two cohort studies. *Human Molecular Genetics, 25*(1), 191–201. https://doi.org/10.1093/hmg/ddv456

Simpkin, A. J., Howe, L. D., Tilling, K., Gaunt, T. R., Lyttleton, O., McArdle, W. L., Ring, S. M., Horvath, S., Smith, G. D., & Relton, C. L. (2017). The epigenetic clock and physical development during childhood and adolescence: Longitudinal analysis from a UK birth cohort. *International Journal of Epidemiology, 46*(2), 549–558. https://doi.org/10.1093/ije/dyw307

Södersten, E., Toskas, K., Rraklli, V., Tiklova, K., Björklund, Å. K., Ringnér, M., Perlmann, T., & Holmberg, J. (2018). A comprehensive map coupling histone modifications with gene regulation in adult dopaminergic and serotonergic neurons. *Nature Communications, 9*(1), Article 1. https://doi.org/10.1038/s41467-018-03538-9

Teschendorff, A. E., & Zheng, S. C. (2017). Cell-type deconvolution in epigenome-wide association studies: A review and recommendations. *Epigenomics, 9*(5), 757–768. https://doi.org/10.2217/epi-2016-0153

Teschendorff, A. E., Yang, Z., Wong, A., Pipinikas, C. P., Jiao, Y., Jones, A., Anjum, S., Hardy, R., Salvesen, H. B., Thirlwell, C., Janes, S. M., Kuh, D., & Widschwendter, M. (2015). Correlation of smoking-associated DNA methylation changes in buccal cells with DNA methylation changes in epithelial cancer. *JAMA Oncology, 1*(4), 476–485. https://doi.org/10.1001/jamaoncol.2015.1053

Timme, N. M., & Lapish, C. (2018). A tutorial for information theory in neuroscience. *ENeuro, 5*(3). https://doi.org/10.1523/ENEURO.0052-18.2018

Tononi, G. (2004). An information integration theory of consciousness. *BMC Neuroscience, 5*(1), 42. https://doi.org/10.1186/1471-2202-5-42

Tononi, G., Boly, M., Massimini, M., & Koch, C. (2016). Integrated information theory: From consciousness to its physical substrate. *Nature Reviews Neuroscience, 17*(7), Article 7. https://doi.org/10.1038/nrn.2016.44

Vieira, M. S., Goulart, V. A. M., Parreira, R. C., Oliveira-Lima, O. C., Glaser, T., Naaldijk, Y. M., Ferrer, A., Savanur, V. H., Reyes, P. A., Sandiford, O., Rameshwar, P., Ulrich, H., Pinto, M. C. X., & Resende, R. R. (2019). Decoding epigenetic cell signaling in neuronal differentiation. *Seminars in Cell & Developmental Biology, 95*, 12–24. https://doi.org/10.1016/j.semcdb.2018.12.006

Viitaniemi, H. M., Verhagen, I., Visser, M. E., Honkela, A., van Oers, K., & Husby, A. (2019). Seasonal variation in genome-wide DNA methylation patterns and the onset of seasonal timing of reproduction in great tits. *Genome Biology and Evolution, 11*(3), 970–983. https://doi.org/10.1093/gbe/evz044

von Uexküll, J. (1926). *Theoretical biology*. K. Paul, Trench, Trubner & Co. Ltd.

von Uexküll, T. (1992). Introduction: The sign theory of Jakob von Uexküll. *Semiotica, 89*(4), 279–316. https://doi.org/10.1515/semi.1992.89.4.279

von Uexküll, J. (2010). *A foray into the worlds of animals and humans: With a theory of meaning* (1st University of Minnesota Press ed). University of Minnesota Press. http://site.ebrary.com/id/10442224

Wagner, A. (1999). Causality in complex systems. *Biology and Philosophy, 14*(1), 83–101. https://doi.org/10.1023/A:1006580900476

Walter, J., & Schickl, H. (Eds.). (2019). *Single-cell analysis in research and medicine. Report of the Interdisciplinary Research Group Gene Technology Report*. Berlin-Brandenburg Academy of Sciences and Humanities.

Wang, X., & Moazed, D. (2017). DNA sequence-dependent epigenetic inheritance of gene silencing and histone H3K9 methylation. *Science, 356*(6333), 88–91. https://doi.org/10.1126/science.aaj2114

Wang, Z., Tang, B., He, Y., & Jin, P. (2016). DNA methylation dynamics in neurogenesis. *Epigenomics, 8*(3), 401–414. https://doi.org/10.2217/epi.15.119

Wang, T., Tsui, B., Kreisberg, J. F., Robertson, N. A., Gross, A. M., Yu, M. K., Carter, H., Brown-Borg, H. M., Adams, P. D., & Ideker, T. (2017). Epigenetic aging signatures in mice livers are slowed by dwarfism, calorie restriction and rapamycin treatment. *Genome Biology, 18*(1), 57. https://doi.org/10.1186/s13059-017-1186-2

Webster, A. P., Plant, D., Ecker, S., Zufferey, F., Bell, J. T., Feber, A., Paul, D. S., Beck, S., Barton, A., Williams, F. M. K., & Worthington, J. (2018). Increased DNA methylation variability in rheumatoid arthritis-discordant monozygotic twins. *Genome Medicine, 10*(1), 64. https://doi.org/10.1186/s13073-018-0575-9

Weidner, C. I., Lin, Q., Koch, C. M., Eisele, L., Beier, F., Ziegler, P., Bauerschlag, D. O., Jöckel, K.-H., Erbel, R., Mühleisen, T. W., Zenke, M., Brümmendorf, T. H., & Wagner, W. (2014). Aging of blood can be tracked by DNA methylation changes at just three CpG sites. *Genome Biology, 15*(2), R24. https://doi.org/10.1186/gb-2014-15-2-r24

Wilson, M. (2002). Six views of embodied cognition. *Psychonomic Bulletin & Review, 9*(4), 625–636. https://doi.org/10.3758/BF03196322

Winograd, S., & Cowan, J. D. (1963). *Reliable computation in the presence of noise* (X956.88). MIT Press; Computer History Museum.

Wright, B. E. (1979). Causality in biological systems. *Trends in Biochemical Sciences, 4*(5), N110–N111. https://doi.org/10.1016/0968-0004(79)90388-8

Xia, L., Ma, S., Zhang, Y., Wang, T., Zhou, M., Wang, Z., & Zhang, J. (2015). Daily variation in global and local DNA methylation in mouse livers. *PLoS One, 10*(2), e0118101. https://doi.org/10.1371/journal.pone.0118101

Yang, Z., Wong, A., Kuh, D., Paul, D. S., Rakyan, V. K., Leslie, R. D., Zheng, S. C., Widschwendter, M., Beck, S., & Teschendorff, A. E. (2016). Correlation of an epigenetic mitotic clock with cancer risk. *Genome Biology, 17*(1), 205. https://doi.org/10.1186/s13059-016-1064-3

Zbieć-Piekarska, R., Spólnicka, M., Kupiec, T., Parys-Proszek, A., Makowska, Ż., Pałeczka, A., Kucharczyk, K., Płoski, R., & Branicki, W. (2015). Development of a forensically useful age prediction method based on DNA methylation analysis. *Forensic Science International. Genetics, 17*, 173–179. https://doi.org/10.1016/j.fsigen.2015.05.001

Zhang, Q., Vallerga, C. L., Walker, R. M., Lin, T., Henders, A. K., Montgomery, G. W., He, J., Fan, D., Fowdar, J., Kennedy, M., Pitcher, T., Pearson, J., Halliday, G., Kwok, J. B., Hickie, I., Lewis, S., Anderson, T., Silburn, P. A., Mellick, G. D., et al. (2019). Improved precision of epigenetic clock estimates across tissues and its implication for biological ageing. *Genome Medicine, 11*(1), 54. https://doi.org/10.1186/s13073-019-0667-1

Chapter 6
The Neuron and the Psyche

It is obvious, too, that such a form of psychology, which has been turned into hypothetical brain-mechanics, can never be of any service as a basis for the mental sciences. (Wundt, 1897, p. 18)

Physiological interpretation must be based, first of all, upon the manifestations of function, and these can be brought, later on, into relation to the anatomical facts. The opposite plan, of erecting elaborate physiological—not to say psychological—hypotheses upon pure anatomical foundations, is, of course, to be rejected without further argument. (Wundt, 1904, p. 50)

In his book *Principles of Physiological Psychology*, Wilhelm Wundt discusses at length how a psychological perspective relates to, but also differs from, a physiological perspective, with the latter being occupied with the analysis and description of the material basis of consciousness according to physical laws and the former representing a distinct mode of description of consciousness based on introspection (1904, pp. 1–15, 50–51). While Wundt clearly stresses that the physiological description of the nervous system cannot replace the psychological perspective at any point, he understands the former as the material basis of the latter and the task of Physiological Psychology to align both perspectives in the long run. Among the different neurophysiological findings and theories that he discussed, he also referenced the neuron doctrine and the idea that neurons are the basic functional units of the nervous system, which were still debated at the time. He writes:

This hypothesis, it is needless to say, ascribes a greatly added importance to the nerve-cell. According to it, the functions of the nervous system are conditioned upon the spheres of function of the individual cells—the "cell" in this sense including as an essential constituent the fibrillar elements issuing from the cell-body. We may therefore regard the nerve-cell together with its processes as the morphological, and presumably also as the functional unit, to which we are in the last resort referred for an understanding of the entire nervous system. (Wundt, 1904, p. 48)

As we see in this founding text of psychological science, ascribing a special role to the single neuron has been part of the discipline's core assumptions from the beginning.

© The Author(s), under exclusive license to Springer Nature Switzerland AG 2024   131
V. Lux, *The Neuron in Context*, SpringerBriefs in Psychology,
https://doi.org/10.1007/978-3-031-55229-8_6

However, also from the beginning, it was emphasized that psychology provides a different perspective not captured by our neurophysiological knowledge of brain cells. Throughout his work, Wundt struggled to further specify the psychological approach and how it relates to physiological, philosophical, and cultural-historical perspectives on the psyche (Mischel, 1970).[1] The distinct object of psychological theories afforded a definition of the psyche that captures its function and is still relatable to these other modes of description.

Using an evolutionary and cultural-historical perspective for the same attempt, Aleksei N. Leontiev defined the psyche in its most basic function as an internal reflection device, which enables the organism to interact with its environment in a goal-directed manner. From this starting point, he identified irritability, exhibited, for example, by protozoa, as the evolutionary preform of basic psychological capacities (Leontyev, 1981). An organism is irritable when a change in the environment that is directly relevant to its metabolism, such as a temperature change or a change in chemical gradients or food availability, functions as a signal for the organism to alter its relationship with the environment, for example, by moving closer or away from the signal, changing its cellular or bodily constitution, or even initiating reproduction. For Leontiev, irritability is still *pre*-psychic and, according to Wundt's categorization, could be fully described in physiological terms. Psychological functions, on the other hand, start with "sensitivity" or the "capacity for sensation," which Leontiev defines as "irritability in relation to that kind of environmental influence that orients the organism by performing a signaling function" (Leontiev, 1981, p. 42). Thus, in the case of sensitivity, the signal according to which an organism initiates a certain activity or behavior is not directly physically relevant to the organism. It only indirectly signals the presence of a relevant environmental condition, such as the shape, color, or odor of food, rather than the nutrients themselves. This indirect connection, however, introduces a certain independence but also potential incongruency in the organism-environment relationship. The indirect signal does not create harm or pleasure and is not agreeable or disagreeable to the organism on its own. It only signals the potential for a change in environmental conditions that might result in harm or pleasure, signaling a threat, food, or mating partner. This has the clear advantage that, in the case of a harmful change in the environment, an organism can direct itself away from the condition without experiencing harm. However, the indirect nature necessitates that the individual *learns* to connect the signal and what it stands for in its metabolic result and *remember* its meaning.

---

[1] Sigrid Weigel (Weigel, 2016, pp. 52–53) points out that Sigmund Freud struggled with the same issues in his early works, as is specifically visible in his text *Entwurf einer Psychologie* (*Project for a Scientific Psychology*, Freud, 1895). In this text, he referred to the "neurons" as "the material particles" related to psychic functions (Freud, 1895, p. 295). Freud ultimately abandoned the attempt to align psychic functions to brain units and the brain's overall anatomical structure but kept the emphasis on the energetic component of psychological functions and how this connects the psyche with the organismic needs and functions of the individual (for example, in his drive theory).

This learned and experience-based connection is what Leontiev defines as the most basic level of psychological function. With this definition, he put the disconnection between the effect on the organism and the inner representation of an environmental condition as the starting point of the evolution of the mind. The gap forces the organism to actively connect internal and external signals by its mental activity, and the brain has evolved as a specialized cell ensemble and organ to serve this purpose. The neuron, in its role as an excitable cell, has been part of this cell ensemble from the beginning. Indeterminacy and plasticity at the single neuron level provide the developmental openness necessary for complex psychological functions to evolve.

I started in the first chapter with a call for a new neuron theory. One that goes beyond the one-directional and atomistic view of the neuron-psyche relationship and accounts for the bidirectional nature and context-dependence of the single neuron. The neuron doctrine constituted the neuron as the basic functional unit of the nervous system, which led to the endeavor to match neural activity at the single-neuron level to psychological function. However, by isolating the neuron from its local and temporal context, we repeatedly fail to match specific single neurons to specific psychological functions. Neuroscience, especially in fields concerned with psychological function, consciousness, and mental health, has already moved away from the single-neuron perspective. However, by focusing only on the systemic and network characteristics of neural activity, we miss out on its cellular and ultimately material basis. Instead of simply discharging the neuron from neuroscience, or replacing it with glia cells, as others have argued (see Chap. 1), we need a neuron theory that captures the role of the neuron in neural activity and ultimately psychological function in its full indeterminacy and fuzziness as it emerges from our growing knowledge about this excitable cell. The plastic and systemic nature of the workings of the brain and its material basis, all the way down to individual brain cells and their molecular components, clearly shows the need for such a conceptual shift.

The Neuron's Power: Combining Indeterminacy with Stabilization of Function

However, emphasizing the neuron's plasticity and functional flexibility is only one side of this endeavor for a new neuron theory. We also need to account for the resistance and inertness of the nervous system, which come with the materiality of the cellular structure of the neuron and the tissue it is embedded in. From the perspective of the neuron, maintaining its cellular integrity and stabilizing its cellular function while still reliably contributing to the neural network and higher functional levels is the actual task. At first glance, the growing evidence of the neuron's flexibility and indeterminacy seems to contradict the stabilization of function, yet it turns out to be its precondition, and vice versa.

This corresponds to the functional affordances and the role of the neuron within the material basis of the psyche. Following Leontiev's definition of the psyche, the indirect nature of the connection between psychological functions and their physiological basis not only leaves room for errors in the representation of the environment. It also comes with the need for some long-term representation of the environmental signal and its meaning beyond its immediate presence. However, neurons do not store the objects, thoughts, and feelings we encounter during our lifetime in the way we would store paper files in an archive or pictures on a flash drive, file by file, or bit by bit. Our brain is neither a library nor a museum nor a computer. It does not catalog the exact firing rate of the five neurons located in the posterior ventral occipitotemporal cortex processing the letter A read in a book on a sunny morning in September with the smell of lemongrass tea in the background. Even if we can specify the neural networks and single-cell activities involved in the perception of a moment full of sensory experiences, there will be enormous differences at the single neuron level every time we perceive, sense, remember, or imagine a similar situation.

This does not mean that neural connections change continuously. As discussed earlier, it is well established that single neurons specialize to serve a specific function as a result of their development in a specific network and cellular context, and that these networks are shaped by our experiences and shape them over the lifespan. It just means that there is some indeterminacy in the relationship between neural activity and psychological functions, and this indeterminacy is not only present at the higher and intermediate levels due to complex interactions but also starts right at the single neuron level. Most importantly, the power of our mental capacities stems from this indeterminacy of the neuron. It allows the single neuron to participate in multiple functional contexts, and, at the same time, first and foremost survive as single cell within the constant stream of neural and molecular information passing through the brain. Multiple complex adaptation processes dealing with this stream of information, integrating and balancing internal and external signals, result in constantly changing neural activity patterns that reflect our internal state in relation to our social-cultural environment. We are even able to consciously switch between the levels of precision of this relationship by intentionally focusing our attention on either the details or the whole picture of our actions and sensations, or some degree of resolution in between. This indicates how strongly the activation of a specific neural network or population of neurons is driven by actions and sensations at the sensory and motor level, by emotional and cognitive processing, and by our overall mental state. However, whether a specific single neuron participates in a neural representation at a specific moment depends on its individual activation state, which results from its developmental and activation history, as well as its immediate cellular environment. Then, again, this immediate cellular and neural network context of a single neuron is shaped by its activity in a recursive homeorhetic manner. A neuron forms and is formed by its context. It is constituted by the neuron's past and present neural network connections and its brain tissue context; ultimately, it is shaped by the social-cultural ecological niche of the individual in which the neuron co-developed as part of the material basis of specific psychological functions. The power of the neuron lies exactly in its ability to integrate both affordances, combining indeterminacy with the stabilization of function in its role as a product and producer of neural activity.

Elements of a New Neuron Theory

Thus, the question of how we conceptualize this indeterminacy of the psyche-brain relationship at the single-neuron level and still capture the neuron as a material and cellular unit with stabilized functionality needs to be at the core of our new neuron theory. A few general assumptions already take shape in light of the findings discussed in the previous chapters.

Accordingly, a new neuron theory must acknowledge the neuron as a cellular unit that is vitally dependent on its cellular, neural network, and bodily context. It needs to account for the fascinating parallelism of highly functional specialization and, at the same time, the broad multifunctional capacity of neurons. It needs to capture that the participation of a neuron in a specific function can be temporary and therefore does not always fully differentiate a single neuron from other neurons, and that maintaining its individual cellular functionality might even stand in contrast to a neuron's participation in neural network activity related to motor and sensory activity or higher psychological functions. We need a neuron theory that reconsolidates homeostatic and synaptic plasticity and their potentially contradictory effects on the neuron's activity state and range, and that models the single neuron as produced by a developing system. A neuron theory that captures the measured neural activity in its systemic character as produced within and by a mixed cellular context. That acknowledges that the indeterminacy of the single neuron–function relationship forces us to reframe the interpretation of single-cell measurements, especially at the molecular level, but that also enables us to account for the cellular materiality of neural activity. This new neuron theory ultimately needs to make thinkable that the neuron is the product and producer of neural activity, formed by and recursively forming its context, functioning as a bidirectional hub and multilevel transmission unit between its connections (Box 6.1).

Box 6.1 Elements of a New Neuron Theory
A new neuron theory needs to capture the role of the neuron in neural activity and psychological functions in its full indeterminacy and fuzziness as it emerges from our knowledge about this excitable cell. This theory needs to align with the following characteristics:

- The neuron is product and producer of neural activity. It is formed by and recursively forming its context.
- As a cellular unit, it vitally depends on its cellular, neural network, and bodily context. Its anatomy, physiology, and functional capacity is produced and maintained by a developing homeorhetic system.

(continued)

Box 6.1 (continued)

- Measured neural activity captures the neuron in its systemic character, produced within and by a mixed cellular context.
- The neuron functions as a bidirectional hub and multi-level transmission unit between its connections.
- The multiple temporal components of the neuron's activity and function contribute to the coordination of its participation in neural network activity and psychological function.
- The neuron can exhibit highly functional specialization and, at the same time, a broad multifunctional capacity. An exact neuron-function relationship is the biological exception in the human brain.
- The structural and functional plasticity range of a neuron is wider than that of other cells. This flexibility needs stabilization to make sense energetically.
- Maintaining its individual cellular functionality can stand in contrast to a neuron's participation in neural network activity, and can lead, for example to contradicting roles of homeostatic plasticity and synaptic plasticity.
- In some circumstances, a neuron's metabolism needs to be stabilized and protected from the constant information influx coming from its immediate environment. Epigenetic mechanisms provide the biophysical properties to contribute to this stabilization in an energy efficient manner.

This new neuron theory must account for the multiple temporal dimensions of the neuron's functionality. As we have seen, functional differentiation and interactive specialization over the life span through progressive integration in variable neural networks, for example, via neural synchronization, strongly shape the developmental and functional context of the single neuron. From such a developmental perspective, neural plasticity can then be understood as the result of change and canalization at different biological levels involved in neural activity, from the neural network to the surrounding cell structure and, most importantly, the glia cells. The feedback loops underlying this canalized plasticity develop throughout the cellular life cycle of the neuron and during interaction with the environment. They represent some form of embodied precondition of a single neuron's activity, with varying degrees of stability and changeability, and depending on the involved functional levels and developmental processes. To determine the degree of indeterminacy at the single neuron level for a specified function, we need to study the processes of canalized stability and plasticity at all functional levels, as well as the feedback loops creating transmission hubs between these levels.

Accounting for the recursive and homeorhetic bidirectionality of a neuron's activity introduces a certain degree of equivocality, indeterminacy, and randomness into our understanding of neural functions. This indeterminacy can be narrowed down to specific, well-characterized functions in a specific individual, but it will persist for the larger task of deciphering the mechanisms of consciousness or comparing the neural basis of psychological functions between individuals. However, this approach is more consistent with the observed biological and biophysical processes. Reaching from higher mental processes to neural networks and glia-neuron communication, the involved levels span a wide spectrum of research fields. This emphasizes the need for an interdisciplinary framework that integrates concepts and methods from different fields.

Finally, part of this reconceptualization of the neuron also needs to be that we account for the brain as a tissue and organ again, instead of solely focusing on single neurons and their neural connections. Nobody would study a single skin or heart cell to understand the function of the skin or the workings of the heart, but with the single neuron in the brain, we simply assume that it represents our mind in miniature. Similar to every other organ that consists of multiple cells, we also need to account for the tissue and organ characteristics of the brain and how this affects the single neuron. We already use tissue parameters, such as cell type composition, tissue-wide mRNA expression measures, proteome quantification, tissue-wide levels of hormones and neurotransmitters, and bulk-tissue neuroepigenetic markers to characterize neural activity. Instead of interpreting these measures as biomarkers for single neuron function or neural network activity, we need to re-evaluate them as biomarkers for full brain tissue. We also need to step away from the abstract notion of the neural basis of psychological function ultimately residing in a single cell, and embrace the imprecision of the involved biological mechanism within the whole brain tissue.

Taking a tissue perspective such as this also implies a change in the notion of embodiment. In contrast to its use in the embodied cognition framework or in phenomenological accounts of the extended mind, this tissue-based notion of embodiment embraces the full biological and socio-cultural embeddedness of the single neuron in its full materiality. Thus, with this new neuron theory, we need to introduce a conceptual shift according to which we understand the brain as both the place where the most complex network connections between neurons are formed as well as the tissue and organ constituted by a functionally fuzzy cell mixture. When interpreted in this sense, the different markers at the molecular level, such as the previously discussed neuroepigenetic data, contribute to the characterization of not only the single neuron and its activation history but, most importantly, the neuron in context.

Reconnecting the Neuron and the Psyche: Developmental Embodiment Research

Based on this contextualized notion of the single neuron, we can no longer assume that psychological function originates from single neuron activity. We must acknowledge that the relationship between a single neuron and psychological function is the result of a co-developmental process to which all functional levels of a developmental system contribute. A thought that has a long tradition in developmental psychology (for a historical overview and discussion, see, for example, Ittel

& Kretschmer, 2007). To reconnect the neuron to the psyche under these assumptions, we need a framework that spans across the involved levels but allows for different mechanisms and processes at different functional levels of the developing system.

Developmental Embodiment Research provides a framework that explicitly models this multilevel character across the contributing socio-cultural, psychological, and biological levels (Krüger & Lux, 2023; Lux et al., 2021). The framework integrates findings in embodiment research across a wide range of different biological (genetic/epigenetic, metabolic, physiological feedback, neural), psychological (sensory/motor activity, sensations, interoception, behavior/action, complex psychological functions, psychiatric symptoms), and social-cultural interaction levels (Lux et al., 2021; see Fig. 6.1). Within this interdisciplinary framework, the embodiment of information and the mobilization of embodied information are understood to be connected via dynamic feedback cycles, spiraling forward in time along the course of development over the lifespan.

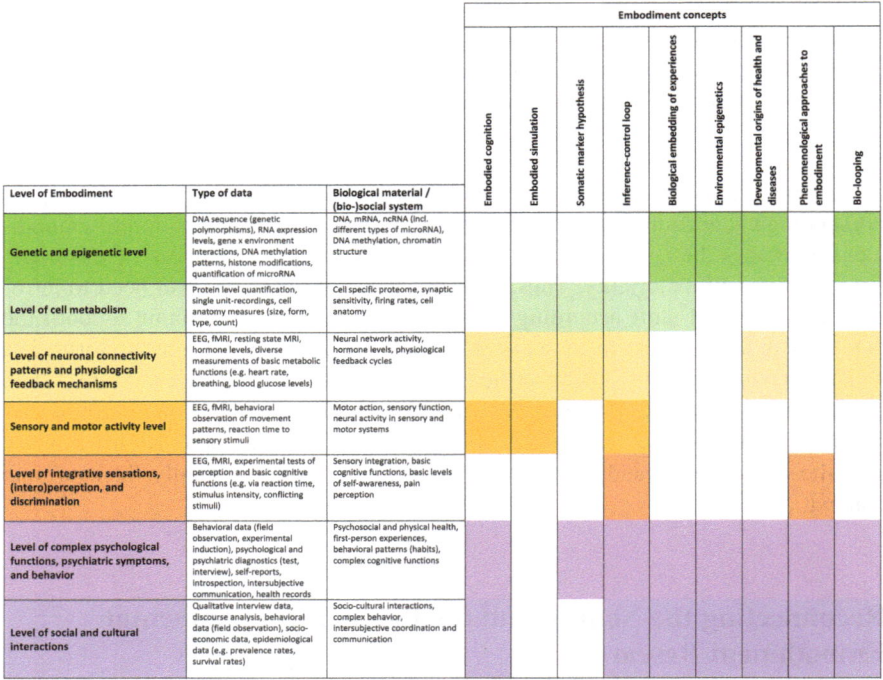

Fig. 6.1 Developmental embodiment research: Functional levels and corresponding research approaches. The figure shows the different functional levels and research approaches integrated within the Developmental Embodiment Research framework. Each color highlights a different functional level of embodiment. For each functional level, the type of data, the addressed biological material or (bio-) social system, and the research approach as represented by a specific embodiment concept are listed. Blank spaces indicate conceptual and methodological gaps between research approaches within a certain level (left to right) and between functional levels within a specific research approach (top to down). (Adapted from Lux et al. (2021), CC BY-NC 4.0 Deed)

Other theoretical approaches that model human development as a multilevel process, such as Developmental Science (Scheithauer et al., 2009), Developmental Systems Theory (Griffiths & Hochman, 2015; Oyama, 2000), and ecological systems approaches (Barrett, 2009), also account for dynamic feedback cycles between different biological and psychosocial levels. Developmental Embodiment Research differs from these approaches in its attempt to provide a meta-framework for level-specific developmental theories. This entails, for example, a broad interdisciplinary perspective integrating research in robotics, computational psychosomatics, social studies, rehabilitation, and movement studies with psychology, psychiatry, neuroscience, developmental biology, genetics, and epigenetics. Most importantly, it focuses on the underlying embodiment processes and the characterization of the different functional levels by the methods and types of data used to identify mechanisms, correlates, and traces of these embodiment processes across these disciplines. The method focus implies that the systemic dynamics are modeled as level-specific at different functional levels.

Consequently, Developmental Embodiment Research implies a multimethod approach and accounts for the different research perspectives through which data are acquired across individual functional levels. Integrating these different research perspectives enables and affords the translation of level-specific data modalities ranging, for example, from molecular data to EEG data to third-person accounts of behavioral data and first-person accounts of subjective experiences. It also implicates that for some instances the translation between levels will not be possible or incomplete, for example, when visual sensory experiences reported from a first-person perspective are studied in their relationship to neural activity measures collected via fMRI or EEG and behavioral data observed from a third-person perspective. However, by differentiating the levels along the applied methods, we can account for method-specific constraints and epistemological gaps between research approaches. This is especially important for questions concerning the mind-brain relationship, where, for example, subjective experiences recollected from a first-person perspective need to be related to different biological levels. Nevertheless, it also applies to translation processes between more closely related functional levels such as those necessary within multi-omics approaches. In their attempt to connect genomics and epigenomics data with proteomics data and different neuronal cell functions, these approaches also face constraints when attempting to relate data across levels. This can be due to methods that cannot be applied to the same cellular material or a lack of knowledge about the interactions of different parts of the observed subsystems within and across levels, as in the case of miRNA and their role in gene expression and protein synthesis.

Furthermore, Developmental Embodiment Research encourages the relationship between the data retrieved at each level to be modeled individually for each psychological function, at least in the first step, especially with regard to the timeline of level interactions. This is motivated by findings in developmental psychology that have demonstrated that there are different developmental trajectories for different psychological functions and their related neural networks (for a more detailed discussion, see Lux et al., 2021). This function-specific approach can account for potential differences in level-specific representations of a specific function, including situations where a function does not impose changes or is not represented at every functional level, but only in a subset of them. For example, while changes are present at the neural activity and subjective experience levels, they might not be reflected in acute measures of hormonal parameters or gene expression patterns. Thus, the approach accounts for functionally specific representations of embodied information, including long-term effects, which are sometimes but not always reflected in acute changes, and short-term effects, which can affect long-term changes at other levels (e.g., in plasticity range or physiological set points) even when they are not maintained at their level of origin, as well as additional systemic buffering mechanisms. Within the framework of Developmental Embodiment Research, each level is considered to exhibit specific timelines and cyclic dynamics along which functional changes are registered, stabilized, or modified. This includes acknowledging a multitude of individual developmental trajectories according to the principle of equifinality, which in its full consequence, warrants a methodological shift from group comparisons to individual case studies.[2]

Using individual case studies to study the basic mechanisms of mind-brain relationships has prominent precedent, be it the determination of the role of the prefrontal cortex in executive functions and its connection with personality as extensively discussed in the historical case of "Phineas Gage" and other patients with similar injuries of the frontal lobes (Mataró et al., 2001; Teles, 2020), or be it the role of emotions in decision-making and the somatic-marker hypothesis put forward by Antonio Damasio and others on the basis of the case study of "Elliot" (Bechara et al., 2000; A. Damasio et al., 2013; A. R. Damasio, 2005). These lesion studies drew theoretical conclusions based on observed functional failure. This is also a promising approach for the study of trans-level interactions in Developmental Embodiment Research (for further discussion, see Krüger & Lux, 2023). The underlying premise is that the functional failure observed in an individual case provides insights into the general mechanisms and interactions that are present in some form in all cases.

Another historical example of the use of individual case studies investigating the physiological foundations of psychological functions, which did not use lesion studies, are the well-known reflex studies in dogs conducted by Ivan P. Pavlov[3] (Chugunow, 2022; Valsiner, 2022b). Pavlov did not study failure of function but the acquisition of a new function through learning–saliva production in response to the bell as a signal for food. He distilled his observations of this conditional reflex into a general theory of psychophysiological plasticity underlying the complex behavior of higher animals and humans. In the analysis of his experimental outcomes, he interpreted differences between the individual dogs' behavioral responses to the

---

[2] For a discussion of such a shift in methodology for psychological research see, for example, Valsiner (2014). Here, it is emphasized that, depending on the research question, such a shift is also necessary for the study of the involved biological processes.

[3] For a conceptual history of the "reflex" in this context and the related laboratory practices, see Vöhringer (2009).

experimental conditions not as measuring error, but as different ranges in the excitability of the nervous system, indicating distinct subtypes (for a detailed discussion and Pavlov's extrapolation to human personality types, see Chugunow, 2022). Valsiner (2022a) describes this method as systemic generalization from the individual case to the generic case and contrasts it to the more frequently used generalization from samples to populations in psychological science. Developmental Embodiment Research builds on this systemic generalization for the translation across functional levels. Within the framework of Developmental Embodiment Research, each individual case is assumed to show differences in the mechanisms contributing to the development of a specific function according to the multiplicity of individual developmental pathways and the principle of equifinality. Consequently, each case adds to the knowledge of the possible level interactions, plasticity ranges, and contributing factors. Overlapping mechanisms between cases then become identifiable as non-specific general mechanisms, while differing mechanisms indicate those specific to a particular individual pathway.

In our effort to conceptualize the neuron in its context, Developmental Embodiment Research thus offers a specific focus that highlights several important characteristics of the neuron and its function. First, the emphasis on embodiment processes captures the materiality of the neuron as a biological cell entity. Second, the combined focus on the embodiment of information and the mobilization of embodied information allows for an understanding of the neuron as a product and producer of neural activity. It qualifies the embodied information related to past and present neural activity as the product and context of the single neuron activity, spanning from neurophysiological and molecular feedback cycles at the cell and tissue levels to larger neural networks, broader physiological feedback processes, and other bodily parameters such as sensory function, muscle constitution, posture, different metabolic parameters, and hormone levels. Third, the analytical distinction of different levels of embodiment at which biological information is present provides a unifying template to further characterize the biological information collected in different research fields according to the methods and the type of data through which it is accessed, measured, and recorded. For brain tissue, this includes epigenetic and gene expression data, proteomic data, neural activity data, and neural connectivity data, either anatomically, in resting state, or in relation to a specific behavior or function. Fourth, the developmental perspective emphasizes that neural function at the single-neuron level depends on the prior developmental and activation histories of the single neuron within its cellular and network context. In addition, the focus on embodiment and its developmental and procedural character corresponds with the concept of canalized plasticity and its homeorhetic character at the single-neuron level, with function being understood as a developed and constantly maintained state in motion. Fifth, the broader context of the neuron is captured by the levels of sensory and motor activity, integrative sensations, complex psychological functions, and social-cultural interactions (see Fig. 6.1).

With its developmental and process-oriented approach and its emphasis on biological embeddedness and embodiment of psychological functions, Developmental Embodiment Research also gives a specific answer to the problem of determining the indeterminacy in the relationship between the single neuron and psychological function. It shifts our focus away from linear causality across all levels, often modeled in statistical terms, to the fuzzy translation processes between specific levels that contribute to the specific function or a subsystem of the function under study. As discussed in the previous chapters, some theoretical and computer modeling approaches already account for the dynamic indeterminacy and fuzziness observed within neural networks. These approaches replace the notion of linear causality with the concept of circular causality between the neuron and a specific function and introduce layers of modulation and systemic indeterminacy. We can use them as tools to relate neuronal signatures to psychological function and vice versa in a nonreductive manner. However, a precondition for this translation process is to specify level-specific methodological representations of the embodied information.

For this endeavor, Developmental Embodiment Research provides a framework within which the single neuron and its activity patterns can be represented at different functional levels using different data modalities, making the transmission between levels the focus point for further investigations. Understanding these translational processes is also crucial to overcoming the level-specific divides currently observed in different fields of neuroscience, from single-cell studies in vitro to network models in silico to data-intensive connectivity research and functional cognitive neuroscience studies. For the specific case of neuroepigenetic data,[4] Developmental Embodiment Research prompts the understanding of the neuroepigenetic trace as tissue-related biomarker of the neuron and its context, which provides an additional layer of information about a neuron's past and present activity state. From this perspective, untwining the time-points at which epigenetic marks are established, reconstructing the corresponding developmental windows, and tracing the epigenetic stabilization cycles will ultimately provide us with hints on a neuron's activation history and canalized plasticity. However, matching such neuroepigenetic profiles with psychological data will likely further unveil the indeterminacy of the neuron-psyche relationship.

The structural and functional plasticity range of a neuron is wider than that of other cells, which is clearly one of its key characteristics. However, from a biological perspective, this flexibility requires stabilization to make sense energetically. What we observe and measure as function of a single neuron is the result of these stabilization processes at a certain time point. Network phenomena, such as synchronicity and their role in integrating neurons in local and distant neural networks, play an important role in these stabilization processes. Also, as discussed, glia-neuron signaling participates in several aspects of these processes. Neuroepigenetic mechanisms have biophysical properties that make them strong contributors to these stabilization processes. We already know that they canalize and enable development at the individual cell level, maintain a specific activation history and state in

---

[4] Of note, the framework of Developmental Embodiment Research also allows to integrate epigenetic data generated from peripheral tissue when studying psychological function. This data then would be integrated according to its functional context within the developmental embodiment process of the specific function under study.

correspondence with structural-genomic epigenetic mechanisms, and coordinate basic transcription loops necessary for the neuron's metabolism and function. In particular, this coordinative role, balancing metabolic and functional affordances, requires further investigation.

A Place for Psychological Theory

In the field of neuroscience, psychological theory was traditionally depicted as fragile and unreliable, and the assumption reigned that once neural function is fully described in biological terms, there is no need for psychological theory anymore. If we could explain all psychological functions, including our subjective experiences, from single neuron cell parameters, we could simply replace psychology with neuroscience. In contrast to this perspective, Wundt (1897, 1904), despite his struggles to carve out the subject of psychology, defined a precise space and methodology for psychological theory, grounded in inner perception via systematic introspection in experimental settings and in relation to physiological, philosophical, and cultural-historical knowledge at the time (Wundt, 1888, 1897; for discussion, see Mischel, 1970). The interpretation of the neuron as the ultimate source of psychological function always threatened to close this space, which would then reduce decisions on motor actions to mere reflections of predetermined neuronal action potentials or learning and memory to changes in synaptic plasticity weights. This placed psychology in the position of an interim discipline to eventually be overcome once neuroscience delivered the formula for the mind-brain relationship.[5]

Having accumulated a tremendous amount of knowledge about the neuron and its workings in the brain throughout the last century, neuroscience is now confronted with the limits of the neuron doctrine. The bottom-up approach from the single neuron to function did not fulfill its promise, and the atomistic foundations of the doctrine have become a hurdle. Simultaneously, top-down approaches, which start with a specific function, gain momentum (for a schematic comparison of bottom-up and top-down approaches, see Fig. 6.2).

Top-down approaches make use of the fine-grained methods available to record neural activity at the single-cell level or to register changes in multiple molecular markers in parallel. Based on this data, they identify with high precision the brain regions or neural networks related to a certain behavioral pattern or higher mental function by analyzing neural activity patterns and shifts in neuronal and molecular biomarkers. However, identified relationships are often only temporary. As discussed for the example of the zebra finch, the recruited neural networks change not only over longer periods, driven by brain development and aging, but also in much shorter timeframes, specifically at the single-neuron level. Moreover, this flexibility at the single-neuron level is likely a core feature of neural representations in the brain.

---

[5] On some of the shortcomings of this view in the more recent debates between psychology and neuroscience see Stam (2015).

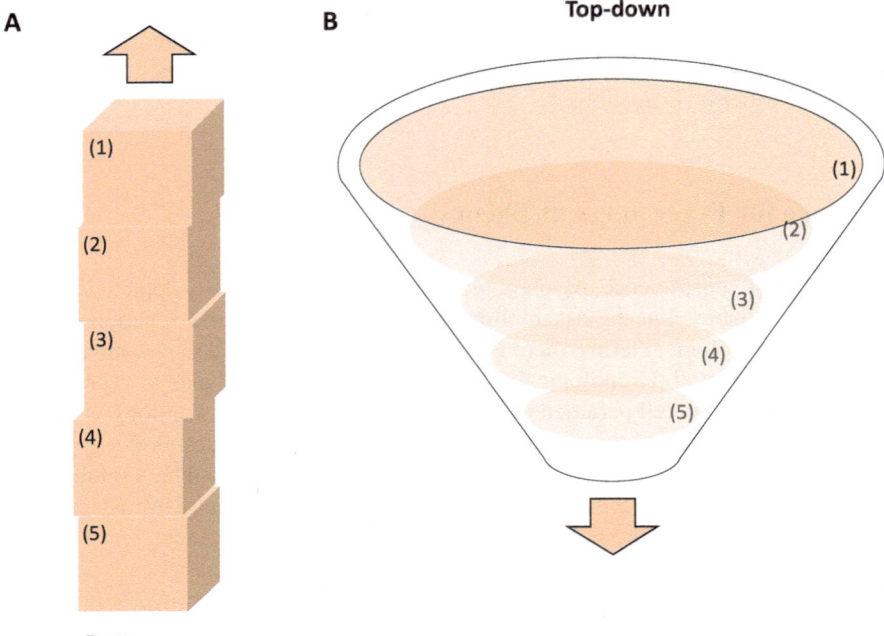

Fig. 6.2 Schematic comparison of bottom-up and top-down approaches to the study of the mind-brain relationship. (**a**) Bottom-up approaches: Knowledge from each area and research field is stacked up on top of each other. Each lower level is a prerequisite for the full understanding of the next higher level. (**b**) Top-down approaches: Knowledge from each area and research field is filtered based on identified functional overlap. Each higher level narrows the relevant information used from the next lower level. Levels of research areas and knowledge production: (1) Psychological function as represented in subjective experiences and behavior (first- and third-person perspective). (2) Specific tasks, behavioral markers, capacities, and experiences contributing to specific psychological functions. (3) Larger neural network dynamics, connectome, activity and temporal coordination of brain areas and functional subunits, etc. (4) Local neural network dynamics, interactions with glia cells, neurotransmitter functions, local brain tissue composition. (5) Neural cell activity, electrophysiological and molecular properties, mechanisms of neural cell differentiation, cell metabolism, etc

On the one hand, this potentially limits the findings of these top-down approaches to rough estimates for specific time points. On the other hand, it shows the power of the reconstructive methods used in top-down analyses to identify neural structures and activity patterns involved in a specific psychological function at a specific point in time. Notably, top-down approaches are already used successfully, for example, in robotics, which operates increasingly from a functional perspective or uses an embodiment approach, starting with the behavior or motor part of the action and then modeling the neural processes necessary for its execution (Cangelosi & Schlesinger, 2015; Eppe et al., 2021; Pfeifer et al., 2007; Vujovic et al., 2017). We also see them contributing to the progress in neurorehabilitation, where usually the function that needs to be restored is well characterized, and the training is directed towards restoring or newly establishing the necessary neural connections in the context of the highly individual patterns of damage and functional failure in individual patients (Grefkes & Fink, 2020; Nshimiyimana et al., 2023; Urendes Jiménez et al., 2014).

This shift, as we see it applied in these practical fields, reframes how neuroscience and psychology relate to each other. When function is the starting point, psychological theory is no longer a placeholder for neuroscience, just describing what emerges from neural activity in different terms, until the precise mechanisms are uncovered. Moreover, the common task is no longer to decode how neurons and other brain cells interact to bring about consciousness or a specific psychological function. Instead, we ask how specific psychological functions recruit groups and networks of neurons, glia cells, etc., to be executed. How the mind uses the brain as an organ to realize its function. This also implies a change in perspective, for example, on functional impairment. In this view, functional impairment is interpreted as a decreased or lost ability to recruit the necessary neural capacities, which could result from individual patterns of cellular damage or lesions but also from dysfunctions at other functional levels, including psychosocial factors. These would then have to be included in the diagnostic process, indicating the need for multi- and cross-level approaches in neuropsychology and neurorehabilitation. In addition, the shift implies at least some conceptualization of the function under study at the psychological or subjective experience level. Accordingly, some neuroscientists have already reached out to phenomenology (Berkovich-Ohana et al., 2020; Varela et al., 2016) and psychoanalysis (Northoff, 2023; Solms, 2018) for this purpose.

Nevertheless, this change in perspective—looking from psychological function to its neural basis instead of the other way around—does not mean that trying to solve the puzzle at the cellular or molecular level does not provide any insights. These studies from below provided us and continue to provide us with important knowledge on how neuronal cells and glia cells develop and function, e.g., where their plasticity range lies, how they interact and depend on each other, under which circumstances and during which timeframes they develop and maintain function, which different connections they build, and which patterns of signal transmission they can exhibit in general. Neuroepigenetic data, for example, can contribute to further characterization of the embodied traces of past and present neural activity patterns related to these basic cell functions. However, activity data and biomarkers will not provide a blueprint on how to build an artificial brain that is more than a basic network of cells. They will help us to understand the potentialities and limits at these lower levels, but they will not be able to determine how these bring about the actual function or behavior. The indeterminacy at the single neuron level constitutes an epistemological gap between the workings of the brain and those of the mind when looking from below. In contrast, if we look from the top, from a specific function as described in psychological terms, this randomness can be narrowed down in retrospective analyses of the recorded data. After a function is executed and can be observed or recollected, we can reconstruct the participation of specific groups and types of neurons, sometimes even down to the single cell, if its activity was recorded, and the contributing molecular mechanisms, if we were able to register them. However, we cannot assume that a particular single cell will participate in this specific function every time it is executed in the future. The indeterminacy at the single-neuron level clearly indicates that neural networks do not function in the way we construct mechanical machines, where every part has a fixed role in the chain of events.

Taken together, both research perspectives need each other. Psychological theory, on the one hand, has to deliver precise descriptions of specific functions, including their functional variation across different timelines and situations, as well as between individuals, in accordance with phenomenological accounts of our subjective experiences. It also has to outline the links to sensory and motor components of the function and the impact of social-cultural contexts with their conditions and affordances. Thus, psychological theory provides the roadmap for which functions or sets of functions we investigate the neurophysiological and cellular basis. Neurological and cell research in neuroscience, on the other hand, provides us with the potentialities of neural function, but also with its cellular and tissue-bound limits. This helps us to understand and calculate how indeterminacy is canalized at the single neuron and network level, how neurons are recruited into specific networks in different brain areas, how and at which timelines they specialize to a certain degree that is not reversible, and where a potential threshold for possible functional changes lies. At what point, degree, or amount of damaged tissue the cellular basis reaches its limits and function will likely be disturbed or not restorable in a particular brain region or at a certain functional level. However, as biological processes are in constant flux, undergoing dynamic changes, and integrated in multiple feedback cycles, function might be restored in a slightly dysfunctional variant, with gaps in precision, by shifting to a higher level, using neural networks within a different brain region, or even using other aspects of our whole bodily constitution and entanglement with the world, including different types of prostheses such as artificial limbs, glasses, and cochlear implants. Finally, we need to assume that shifts in the neural structures contributing to a specific psychological function, or even a simple motor action, occur daily within all of us, with the result that the neural basis of our mental capacities changes constantly.

To integrate both perspectives, we need an interdisciplinary approach that integrates psychological and neurophysiological knowledge in the study of the mind-brain relationship. Instead of simply treating them as different perspectives on the same subject, as Wundt did when he advocated for a proper space of psychology theory, we need to reframe them as parts of the overall puzzle to which other types of knowledge also contribute, including anatomy, body physiology, evolutionary theory, developmental biology, cultural-historical knowledge, and social theories. Developmental Embodiment Research provides an interdisciplinary platform for this program. It sets a multilevel framework with a developmental perspective, which details a dedicated space for psychological theory, among other disciplinary approaches, studying the developing mind and brain over the lifespan. Within all of this, the neuron still holds a central position as an exciting and excitable cell, yet in its double role, as a product and producer of neural activity, embedded in its cellular and wider functional context, and co-developing within the full organism-environment system, in its capacity as a biological entity and, at the same time, as socioculturally produced scientific fact. In short, as a neuron in context.

References

Barrett, L. F. (2009). The future of psychology: Connecting mind to brain. *Perspectives on Psychological Science: A Journal of the Association for Psychological Science, 4*(4), 326–339. https://doi.org/10.1111/j.1745-6924.2009.01134.x

Bechara, A., Damasio, H., & Damasio, A. R. (2000). Emotion, decision making and the orbitofrontal cortex. *Cerebral Cortex (New York, N.Y.: 1991), 10*(3), 295–307. https://doi.org/10.1093/cercor/10.3.295

Berkovich-Ohana, A., Dor-Ziderman, Y., Trautwein, F.-M., Schweitzer, Y., Nave, O., Fulder, S., & Ataria, Y. (2020). The Hitchhiker's guide to neurophenomenology – The case of studying self boundaries with meditators. *Frontiers in Psychology, 11*, 1680. https://www.frontiersin.org/articles/10.3389/fpsyg.2020.01680

Cangelosi, A., & Schlesinger, M. (2015). *Developmental robotics: From babies to robots.* https://doi.org/10.7551/mitpress/9320.001.0001

Chugunow, L. (2022). Ivan P. Pavlov's dedication to science: Investigating the behavior and the human psyche. In J. Valsiner (Ed.), *One dog is enough. Ivan P. Pavlov's contribution to ideographic science* (pp. 1–45). Information Age Publishing.

Damasio, A. R. (2005). *Descartes' error: Emotion, reason and the human brain.* Penguin Books.

Damasio, A., Damasio, H., & Tranel, D. (2013). Persistence of feelings and sentience after bilateral damage of the insula. *Cerebral Cortex (New York, N.Y.: 1991), 23*(4), 833–846. https://doi.org/10.1093/cercor/bhs077

Eppe, M., Wermter, S., Hafner, V. V., & Nagai, Y. (2021). Developmental robotics and its role towards artificial general intelligence. *KI – Künstliche Intelligenz, 35*(1), 5–7. https://doi.org/10.1007/s13218-021-00706-w

Freud, S. (1895). A project for a scientific psychology. In *Standard edition* (Vol. I, pp. 283–397). Hogarth.

Grefkes, C., & Fink, G. R. (2020). Recovery from stroke: Current concepts and future perspectives. *Neurological Research and Practice, 2*(1), 17. https://doi.org/10.1186/s42466-020-00060-6

Griffiths, P. E., & Hochman, A. (2015). Developmental systems theory. In *ELS* (pp. 1–7). Wiley. https://doi.org/10.1002/9780470015902.a0003452.pub2

Ittel, A., & Kretschmer, T. (2007). Historical roots of developmental science. *International Journal of Developmental Science, 1*(1), 23–32. https://doi.org/10.3233/DEV-2007-1103

Krüger, M., & Lux, V. (2023). Failure of motor function—A Developmental Embodiment Research perspective on the systemic effects of stress. *Frontiers in Human Neuroscience, 17*, 1083200. https://www.frontiersin.org/articles/10.3389/fnhum.2023.1083200

Leontiev, A. N. (1981). *Problems of the development of the mind.* Progress Publ.

Lux, V., Non, A. L., Pexman, P. M., Stadler, W., Weber, L. A. E., & Krüger, M. (2021). A developmental framework for embodiment research: The next step toward integrating concepts and methods. *Frontiers in Systems Neuroscience, 15*, 672740. https://doi.org/10.3389/fnsys.2021.672740

Mataró, M., Jurado, M. A., García-Sánchez, C., Barraquer, L., Costa-Jussà, F. R., & Junqué, C. (2001). Long-term effects of bilateral frontal brain lesion: 60 years after injury with an iron bar. *Archives of Neurology, 58*(7), 1139–1142. https://doi.org/10.1001/archneur.58.7.1139

Mischel, T. (1970). Wundt and the conceptual foundations of psychology. *Philosophy and Phenomenological Research, 31*(1), 1–26. https://doi.org/10.2307/2105977

Northoff, G. (2023). *Neuropsychoanalysis: A contemporary introduction.* Routledge.

Nshimiyimana, J., Uwihoreye, P., Muhigirwa, J. C., Niyonsega, T., Nshimiyimana, J., Uwihoreye, P., Muhigirwa, J. C., & Niyonsega, T. (2023). Neurofunctional intervention approaches. In *Neurorehabilitation and physical therapy.* IntechOpen. https://doi.org/10.5772/intechopen.106604

Oyama, S. (2000). *The ontogeny of information: Developmental systems and evolution* (2nd ed., rev. and expanded ed.). Duke University Press.

Pfeifer, R., Lungarella, M., & Iida, F. (2007). Self-organization, embodiment, and biologically inspired robotics. *Science (New York, N.Y.), 318*(5853), 1088–1093. https://doi.org/10.1126/science.1145803

Scheithauer, H., Niebank, K., & Ittel, A. (2009). Developmental science: Integrating knowledge about dynamic processes in human development. In J. Valsiner, P. C. M. Molenaar, M. C. D. P. Lyra, & N. Chaudhary (Eds.), *Dynamic process methodology in the social and developmental sciences* (pp. 595–617). Springer US. https://doi.org/10.1007/978-0-387-95922-1_26

Solms, M. L. (2018). The neurobiological underpinnings of psychoanalytic theory and therapy. *Frontiers in Behavioral Neuroscience, 12*, 294. https://www.frontiersin.org/articles/10.3389/fnbeh.2018.00294

Stam, H. J. (2015). The neurosciences and the search for a unified psychology: The science and esthetics of a single framework. *Frontiers in Psychology, 6*, 1467. https://doi.org/10.3389/fpsyg.2015.01467

Teles, R. V. (2020). Phineas Gage's great legacy. *Dementia & Neuropsychologia, 14*(4), 419–421. https://doi.org/10.1590/1980-57642020dn14-040013

Urendes Jiménez, E., Flores Caballero, A., Molina Rueda, F., Iglesias Giménez, J., & Oboe, R. (2014). Reverse-engineer the brain: Perspectives and challenges. In J. L. Pons & D. Torricelli (Eds.), *Emerging therapies in neurorehabilitation* (pp. 173–188). Springer. https://doi.org/10.1007/978-3-642-38556-8_9

Valsiner, J. (2014). Needed for cultural psychology: Methodology in a new key. *Culture & Psychology, 20*(1), 3–30. https://doi.org/10.1177/1354067X13515941

Valsiner, J. (2022a). Pathways to generalization: General knowledge as abstract complementation. In J. Valsiner (Ed.), *One dog is enough: Ivan P. Pavlov's contribution to ideographic science* (pp. 47–68). Information Age Publishing.

Valsiner, J. (2022b). Return to the dog: Ivan P. Pavlov as a pioneer of idiographic science. In J. Valsiner (Ed.), *One dog is enough: Ivan P. Pavlov's contribution to ideographic science* (pp. vii–xiv). Information Age Publishing.

Varela, F. J., Thompson, E., & Rosch, E. (2016). *The embodied mind: Cognitive science and human experience* (revised ed.). MIT Press.

Vöhringer, M. (2009). Reflex. Begriff und Experiment. In *Begriffsgeschichte der Naturwissenschaften* (pp. 203–214). De Gruyter. https://doi.org/10.1515/9783110213034.3.203

Vujovic, V., Rosendo, A., Brodbeck, L., & Iida, F. (2017). Evolutionary developmental robotics: Improving morphology and control of physical robots. *Artificial Life, 23*(2), 169–185. https://doi.org/10.1162/ARTL_a_00228

Weigel, S. (2016). Embodiment in simulation theory and cultural science, with remarks on the coding-problem of neuroscience. In S. Weigel & G. Scharbert (Eds.), *A neuro-psychoanalytical dialogue for bridging Freud and the neurosciences* (pp. 47–71). Springer International Publishing. https://doi.org/10.1007/978-3-319-17605-5_4

Wundt, W. (1888). Selbstbeobachtung und innere Wahrnehmung. *Philosophische Studien, 4*, 292–309.

Wundt, W. (1897). *Outline of psychology* (pp. xviii, 342). Wilhelm Engelmann. https://doi.org/10.1037/12908-000.

Wundt, W. (1904). *Principles of physiological psychology.* Swan Sonnenschein & Co. Lim.

References

Abu Hamdeh, S., Ciuculete, D.-M., Sarkisyan, D., Bakalkin, G., Ingelsson, M., Schiöth, H. B., & Marklund, N. (2021). Differential DNA methylation of the genes for amyloid precursor protein, tau, and neurofilaments in human traumatic brain injury. *Journal of Neurotrauma, 38*(12), 1679–1688. https://doi.org/10.1089/neu.2020.7283

Aerdker, S., Feng, J., & Schöner, G. (2022). Habituation and dishabituation in motor behavior: Experiment and neural dynamic model. *Frontiers in Psychology, 13*, 717669. https://www.frontiersin.org/articles/10.3389/fpsyg.2022.717669

Agboola, O. S., Hu, X., Shan, Z., Wu, Y., & Lei, L. (2021). Brain organoid: A 3D technology for investigating cellular composition and interactions in human neurological development and disease models in vitro. *Stem Cell Research & Therapy, 12*(1), 430. https://doi.org/10.1186/s13287-021-02369-8

Ahmad, S., Srivastava, R. K., Singh, P., Naik, U. P., & Srivastava, A. K. (2022). Role of extracellular vesicles in glia-neuron intercellular communication. *Frontiers in Molecular Neuroscience, 15*, 844194. https://doi.org/10.3389/fnmol.2022.844194

Allen, N. J., & Lyons, D. A. (2018). Glia as architects of central nervous system formation and function. *Science (New York, N.Y.), 362*(6411), 181. https://doi.org/10.1126/science.aat0473

Altimus, C. M., Marlin, B. J., Charalambakis, N. E., Colón-Rodríguez, A., Glover, E. J., Izbicki, P., Johnson, A., Lourenco, M. V., Makinson, R. A., McQuail, J., Obeso, I., Padilla-Coreano, N., & Wells, M. F. (2020). The next 50 years of neuroscience. *Journal of Neuroscience, 40*(1), 101–106. https://doi.org/10.1523/JNEUROSCI.0744-19.2019

Amiri, A., Coppola, G., Scuderi, S., Wu, F., Roychowdhury, T., Liu, F., Pochareddy, S., Shin, Y., Safi, A., Song, L., Zhu, Y., Sousa, A. M. M., Gerstein, M., Crawford, G. E., Sestan, N., Abyzov, A., Vaccarino, F. M., & PsychENCODE Consortium. (2018). Transcriptome and epigenome landscape of human cortical development modeled in organoids. *Science (New York, N.Y.), 362*(6420), eaat6720. https://doi.org/10.1126/science.aat6720

Arber, S. (2017). Organization and function of neuronal circuits controlling movement. *EMBO Molecular Medicine, 9*(3), 281–284. https://doi.org/10.15252/emmm.201607226

Auksztulewicz, R., Myers, N. E., Schnupp, J. W., & Nobre, A. C. (2019). Rhythmic temporal expectation boosts neural activity by increasing neural gain. *Journal of Neuroscience, 39*(49), 9806–9817. https://doi.org/10.1523/JNEUROSCI.0925-19.2019

Azzi, A., Dallmann, R., Casserly, A., Rehrauer, H., Patrignani, A., Maier, B., Kramer, A., & Brown, S. A. (2014). Circadian behavior is light-reprogrammed by plastic DNA methylation. *Nature Neuroscience, 17*(3), 377–382. https://doi.org/10.1038/nn.3651

© The Editor(s) (if applicable) and The Author(s), under exclusive license to Springer Nature Switzerland AG 2024
V. Lux, *The Neuron in Context*, SpringerBriefs in Psychology, https://doi.org/10.1007/978-3-031-55229-8

Bacci, A., Verderio, C., Pravettoni, E., & Matteoli, M. (1999). The role of glial cells in synaptic function. *Philosophical Transactions of the Royal Society B: Biological Sciences, 354*(1381), 403–409.

Backer, K. C. (2022). Introduction to experimental methods in cognitive neuroscience. In *Mind, cognition, and neuroscience*. Routledge.

Backes, E., & Hemby, S. E. (2003). Discrete cell gene profiling of ventral tegmental dopamine neurons after acute and chronic cocaine self-administration. *The Journal of Pharmacology and Experimental Therapeutics, 307*(2), 450–459. https://doi.org/10.1124/jpet.103.054965

Bagot, R. C., Parise, E. M., Peña, C. J., Zhang, H.-X., Maze, I., Chaudhury, D., Persaud, B., Cachope, R., Bolaños-Guzmán, C. A., Cheer, J., Deisseroth, K., Han, M.-H., & Nestler, E. J. (2015). Ventral hippocampal afferents to the nucleus accumbens regulate susceptibility to depression. *Nature Communications, 6*, 7062. https://doi.org/10.1038/ncomms8062

Bailey, C. H., & Kandel, E. R. (1995). Molecular and structural mechanisms underlying long-term memory. In M. S. Gazzaniga (Ed.), *The cognitive neurosciences* (pp. 19–36). MIT Press.

Bailey, C. H., Bartsch, D., & Kandel, E. R. (1996). Toward a molecular definition of long-term memory storage. *Proceedings of the National Academy of Sciences, 93*(24), 13445–13452. https://doi.org/10.1073/pnas.93.24.13445

Baltes, P. B., Reese, H. W., & Lipsitt, L. P. (1980). Life-span developmental psychology. *Annual Review of Psychology, 31*, 65–110. https://doi.org/10.1146/annurev.ps.31.020180.000433

Banker, G. A. (1980). Trophic interactions between astroglial cells and hippocampal neurons in culture. *Science, 209*(4458), 809–810. https://doi.org/10.1126/science.7403847

Banker, G. A. (2018). The development of neuronal polarity: A retrospective view. *Journal of Neuroscience, 38*(8), 1867–1873. https://doi.org/10.1523/JNEUROSCI.1372-16.2018

Barker, D. J. P. (2007). The origins of the developmental origins theory. *Journal of Internal Medicine, 261*(5), 412–417. https://doi.org/10.1111/j.1365-2796.2007.01809.x

Barr, M. S., Farzan, F., Rusjan, P. M., Chen, R., Fitzgerald, P. B., & Daskalakis, Z. J. (2009). Potentiation of gamma oscillatory activity through repetitive transcranial magnetic stimulation of the dorsolateral prefrontal cortex. *Neuropsychopharmacology, 34*(11), Article 11. https://doi.org/10.1038/npp.2009.79

Barr, M. S., Rajji, T. K., Zomorrodi, R., Radhu, N., George, T. P., Blumberger, D. M., & Daskalakis, Z. J. (2017). Impaired theta-gamma coupling during working memory performance in schizophrenia. *Schizophrenia Research, 189*, 104–110. https://doi.org/10.1016/j.schres.2017.01.044

Barreto, G., Schäfer, A., Marhold, J., Stach, D., Swaminathan, S. K., Handa, V., Döderlein, G., Maltry, N., Wu, W., Lyko, F., & Niehrs, C. (2007). Gadd45a promotes epigenetic gene activation by repair-mediated DNA demethylation. *Nature, 445*(7128), 671–675. https://doi.org/10.1038/nature05515

Barrett, L. F. (2009). The future of psychology: Connecting mind to brain. *Perspectives on Psychological Science: A Journal of the Association for Psychological Science, 4*(4), 326–339. https://doi.org/10.1111/j.1745-6924.2009.01134.x

Bartos, M., Vida, I., & Jonas, P. (2007). Synaptic mechanisms of synchronized gamma oscillations in inhibitory interneuron networks. *Nature Reviews. Neuroscience, 8*(1), 45–56. https://doi.org/10.1038/nrn2044

Bassett, D. S., & Bullmore, E. T. (2016). Small-world brain networks revisited. *The Neuroscientist: A Review Journal Bringing Neurobiology, Neurology and Psychiatry, 23*(5), 499–516. https://doi.org/10.1177/1073858416667720

Bateson, P., & Gluckman, P. (2012). Plasticity and robustness in development and evolution. *International Journal of Epidemiology, 41*(1), 219–223. https://doi.org/10.1093/ije/dyr240

Baumbach, J. L., & Zovkic, I. B. (2020). Hormone-epigenome interactions in behavioural regulation. *Hormones and Behavior, 118*, 104680. https://doi.org/10.1016/j.yhbeh.2020.104680

Bayraktar, G., & Kreutz, M. R. (2018). Neuronal DNA methyltransferases: Epigenetic mediators between synaptic activity and gene expression? *The Neuroscientist, 24*(2), 171–185. https://doi.org/10.1177/1073858417707457

Bayraktar, G., Yuanxiang, P., Confettura, A. D., Gomes, G. M., Raza, S. A., Stork, O., Tajima, S., Suetake, I., Karpova, A., Yildirim, F., & Kreutz, M. R. (2020). Synaptic control of

DNA methylation involves activity-dependent degradation of DNMT3A1 in the nucleus. *Neuropsychopharmacology, 45*(12), Article 12. https://doi.org/10.1038/s41386-020-0780-2

Bearer, E. L., & Mulligan, B. S. (2018). Epigenetic changes associated with early life experiences: Saliva, a biospecimen for DNA methylation signatures. *Current Genomics, 19*(8), 676–698. https://doi.org/10.2174/1389202919666180307150508

Bechara, A., Damasio, H., & Damasio, A. R. (2000). Emotion, decision making and the orbitofrontal cortex. *Cerebral Cortex (New York, N.Y.: 1991), 10*(3), 295–307. https://doi.org/10.1093/cercor/10.3.295

Beck, S. (2014). The human epigenome project: Past, present, and future. In *Reference module in biomedical sciences*. Elsevier. https://doi.org/10.1016/B978-0-12-801238-3.00096-9

Bell, C. G., Lowe, R., Adams, P. D., Baccarelli, A. A., Beck, S., Bell, J. T., Christensen, B. C., Gladyshev, V. N., Heijmans, B. T., Horvath, S., Ideker, T., Issa, J.-P. J., Kelsey, K. T., Marioni, R. E., Reik, W., Relton, C. L., Schalkwyk, L. C., Teschendorff, A. E., Wagner, W., et al. (2019). DNA methylation aging clocks: Challenges and recommendations. *Genome Biology, 20*(1), 249. https://doi.org/10.1186/s13059-019-1824-y

Bengoetxea, H., Ortuzar, N., Bulnes, S., Rico-Barrio, I., Lafuente, J. V., & Argandoña, E. G. (2012). Enriched and deprived sensory experience induces structural changes and rewires connectivity during the postnatal development of the brain. *Neural Plasticity, 2012*, e305693. https://doi.org/10.1155/2012/305693

Beniaguev, D., Segev, I., & London, M. (2021). Single cortical neurons as deep artificial neural networks. *Neuron, 109*(17), 2727–2739.e3. https://doi.org/10.1016/j.neuron.2021.07.002

Bentivoglio, M., Cotrufo, T., Ferrari, S., Tesoriero, C., Mariotto, S., Bertini, G., Berzero, A., & Mazzarello, P. (2019). The original histological slides of Camillo Golgi and his discoveries on neuronal structure. *Frontiers in Neuroanatomy, 13*, 3. https://www.frontiersin.org/articles/10.3389/fnana.2019.00003

Bergsma, T., & Rogaeva, E. (2020). DNA methylation clocks and their predictive capacity for aging phenotypes and healthspan. *Neuroscience Insights, 15*, 2633105520942221. https://doi.org/10.1177/2633105520942221

Berkovich-Ohana, A., Dor-Ziderman, Y., Trautwein, F.-M., Schweitzer, Y., Nave, O., Fulder, S., & Ataria, Y. (2020). The Hitchhiker's guide to neurophenomenology – The case of studying self boundaries with meditators. *Frontiers in Psychology, 11*, 1680. https://www.frontiersin.org/articles/10.3389/fpsyg.2020.01680

Berlucchi, G., & Buchtel, H. A. (2009). Neuronal plasticity: Historical roots and evolution of meaning. *Experimental Brain Research, 192*(3), 307–319. https://doi.org/10.1007/s00221-008-1611-6

Bheda, P., & Schneider, R. (2014). Epigenetics reloaded: The single-cell revolution. *Trends in Cell Biology, 24*(11), 712–723. https://doi.org/10.1016/j.tcb.2014.08.010

Bi, G., & Poo, M. (2001). Synaptic modification by correlated activity: Hebb's postulate revisited. *Annual Review of Neuroscience, 24*, 139–166. https://doi.org/10.1146/annurev.neuro.24.1.139

Bialek, W., Rieke, F., de Ruyter van Steveninck, R. R., & Warland, D. (1991). Reading a neural code. *Science (New York, N.Y.), 252*(5014), 1854–1857. https://doi.org/10.1126/science.2063199

Bianco-Miotto, T., Craig, J. M., Gasser, Y. P., van Dijk, S. J., & Ozanne, S. E. (2017). Epigenetics and DOHaD: From basics to birth and beyond. *Journal of Developmental Origins of Health and Disease, 8*(5), 513–519. https://doi.org/10.1017/S2040174417000733

Bibikova, M., Barnes, B., Tsan, C., Ho, V., Klotzle, B., Le, J. M., Delano, D., Zhang, L., Schroth, G. P., Gunderson, K. L., Fan, J.-B., & Shen, R. (2011). High density DNA methylation array with single CpG site resolution. *Genomics, 98*(4), 288–295. https://doi.org/10.1016/j.ygeno.2011.07.007

Binder, M. D., Hirokawa, N., & Windhorst, U. (Eds.). (2009). Neuron. In *Encyclopedia of neuroscience* (p. 2751). Springer. https://doi.org/10.1007/978-3-540-29678-2_3902

Bird, A. (2007). Perceptions of epigenetics. *Nature, 447*(7143), Article 7143. https://doi.org/10.1038/nature05913

Bladon, J. H., Sheehan, D. J., Freitas, C. S. D., & Howard, M. W. (2019). In a temporally segmented experience hippocampal neurons represent temporally drifting context but not discrete segments. *Journal of Neuroscience, 39*(35), 6936–6952. https://doi.org/10.1523/JNEUROSCI.1420-18.2019

Bock, J., Wainstock, T., Braun, K., & Segal, M. (2015). Stress in utero: Prenatal programming of brain plasticity and cognition. *Biological Psychiatry, 78*(5), 315–326. https://doi.org/10.1016/j.biopsych.2015.02.036

Bocklandt, S., Lin, W., Sehl, M. E., Sánchez, F. J., Sinsheimer, J. S., Horvath, S., & Vilain, E. (2011). Epigenetic predictor of age. *PLoS One, 6*(6), e14821. https://doi.org/10.1371/journal.pone.0014821

Bockmühl, Y., Patchev, A. V., Madejska, A., Hoffmann, A., Sousa, J. C., Sousa, N., Holsboer, F., Almeida, O. F. X., & Spengler, D. (2015). Methylation at the CpG island shore region upregulates Nr3c1 promoter activity after early-life stress. *Epigenetics, 10*(3), 247–257. https://doi.org/10.1080/15592294.2015.1017199

Bonini, L., Rozzi, S., Serventi, F. U., Simone, L., Ferrari, P. F., & Fogassi, L. (2010). Ventral premotor and inferior parietal cortices make distinct contribution to action organization and intention understanding. *Cerebral Cortex (New York, N.Y.: 1991), 20*(6), 1372–1385. https://doi.org/10.1093/cercor/bhp200

Borrelli, E., Nestler, E. J., Allis, C. D., & Sassone-Corsi, P. (2008). Decoding the epigenetic language of neuronal plasticity. *Neuron, 60*(6), 961–974. https://doi.org/10.1016/j.neuron.2008.10.012

Borst, A., & Theunissen, F. E. (1999). Information theory and neural coding. *Nature Neuroscience, 2*(11), Article 11. https://doi.org/10.1038/14731

Bota, M., & Swanson, L. W. (2007). The neuron classification problem. *Brain Research Reviews, 56*(1), 79–88. https://doi.org/10.1016/j.brainresrev.2007.05.005

Bouton, M. E., Maren, S., & McNally, G. P. (2021). Behavioral and neurobiological mechanisms of pavlovian and instrumental extinction learning. *Physiological Reviews, 101*(2), 611–681. https://doi.org/10.1152/physrev.00016.2020

Bower, J. M., & Beeman, D. (Eds.). (1998). *The book of GENESIS* (2nd ed.). Springer. https://doi.org/10.1007/978-1-4612-1634-6

Bowlby, J. (1974). *Attachment and loss. Vol. 1: Attachment* (Repr., with corr. – 1974. – XX, 428 S. – (... ; 79)). Hogarth Press.

Bowlby, J. (1988). *A secure base: Parent-child attachment and healthy human development* (pp. xii, 205). Basic Books.

Brägelmann, J., & Lorenzo Bermejo, J. (2018). A comparative analysis of cell-type adjustment methods for epigenome-wide association studies based on simulated and real data sets. *Briefings in Bioinformatics, 20*(6), 2055–2065. https://doi.org/10.1093/bib/bby068

Bramble, M. S., Roach, L., Lipson, A., Vashist, N., Eskin, A., Ngun, T., Gosschalk, J. E., Klein, S., Barseghyan, H., Arboleda, V. A., & Vilain, E. (2016). Sex-specific effects of testosterone on the sexually dimorphic transcriptome and epigenome of embryonic neural stem/progenitor cells. *Scientific Reports, 6*(1), Article 1. https://doi.org/10.1038/srep36916

Brette, R. (2019). Is coding a relevant metaphor for the brain? *Behavioral and Brain Sciences, 42*, e215. https://doi.org/10.1017/S0140525X19000049

Brod, G., Bunge, S. A., & Shing, Y. L. (2017). Does one year of schooling improve children's cognitive control and alter associated brain activation? *Psychological Science, 28*(7), 967–978. https://doi.org/10.1177/0956797617699838

Brown, E. N., Kass, R. E., & Mitra, P. P. (2004). Multiple neural spike train data analysis: State-of-the-art and future challenges. *Nature Neuroscience, 7*(5), Article 5. https://doi.org/10.1038/nn1228

Bruel-Jungerman, E., Rampon, C., & Laroche, S. (2007). Adult hippocampal neurogenesis, synaptic plasticity and memory: Facts and hypotheses. *Reviews in the Neurosciences, 18*(2), 93–114. https://doi.org/10.1515/revneuro.2007.18.2.93

Budd, A. M., Robins, J. B., Whybird, O., & Jerry, D. R. (2022). Epigenetics underpins phenotypic plasticity of protandrous sex change in fish. *Ecology and Evolution, 12*(3), e8730. https://doi.org/10.1002/ece3.8730

Bühler, K. (1960). *Das Gestaltprinzip im Leben des Menschen und der Tiere.* https://ids-pub.bsz-bw.de/frontdoor/index/index/docId/5913

Bühler, K. (1999). *Sprachtheorie: Die Darstellungsfunktion der Sprache* (3. Aufl., ungekürzter Neudr. d. Ausg. Jena, Fischer, 1934). Lucius und Lucius.

Bullock, T. H., Bennett, M. V. L., Johnston, D., Josephson, R., Marder, E., & Fields, R. D. (2005). The neuron doctrine, redux. *Science, 310*(5749), 791. https://doi.org/10.1126/science.1114394

Bunge, S. A., & Leib, E. R. (2020). How does education hone reasoning ability? *Current Directions in Psychological Science, 29*(2), 167–173. https://doi.org/10.1177/0963721419898818

Campbell, K. (2003). Signaling to and from radial glia. *Glia, 43*(1), 44–46. https://doi.org/10.1002/glia.10247

Campbell, R. R., & Wood, M. A. (2019). How the epigenome integrates information and reshapes the synapse. *Nature Reviews. Neuroscience, 20*(3), 133–147. https://doi.org/10.1038/s41583-019-0121-9

Campelo, T., Augusto, E., Chenouard, N., de Miranda, A., Kouskoff, V., Camus, C., Choquet, D., & Gambino, F. (2020). AMPAR-dependent synaptic plasticity initiates cortical remapping and adaptive behaviors during sensory experience. *Cell Reports, 32*(9), 108097. https://doi.org/10.1016/j.celrep.2020.108097

Cangelosi, A., & Schlesinger, M. (2015). *Developmental robotics: From babies to robots.* https://doi.org/10.7551/mitpress/9320.001.0001

Capelli, P., Pivetta, C., Soledad Esposito, M., & Arber, S. (2017). Locomotor speed control circuits in the caudal brainstem. *Nature, 551*(7680), Article 7680. https://doi.org/10.1038/nature24064

Capper, D., Jones, D. T. W., Sill, M., Hovestadt, V., Schrimpf, D., Sturm, D., Koelsche, C., Sahm, F., Chavez, L., Reuss, D. E., Kratz, A., Wefers, A. K., Huang, K., Pajtler, K. W., Schweizer, L., Stichel, D., Olar, A., Engel, N. W., Lindenberg, K., et al. (2018). DNA methylation-based classification of central nervous system tumours. *Nature, 555*(7697), Article 7697. https://doi.org/10.1038/nature26000

Cardin, J. A., Carlén, M., Meletis, K., Knoblich, U., Zhang, F., Deisseroth, K., Tsai, L.-H., & Moore, C. I. (2009). Driving fast-spiking cells induces gamma rhythm and controls sensory responses. *Nature, 459*(7247), Article 7247. https://doi.org/10.1038/nature08002

Chambers, A. R., & Rumpel, S. (2017). A stable brain from unstable components: Emerging concepts and implications for neural computation. *Neuroscience, 357*, 172–184. https://doi.org/10.1016/j.neuroscience.2017.06.005

Chanda, P., Costa, E., Hu, J., Sukumar, S., Van Hemert, J., & Walia, R. (2020). Information theory in computational biology: Where we stand today. *Entropy (Basel, Switzerland), 22*(6), 627. https://doi.org/10.3390/e22060627

Chatterjee, S., Mizar, P., Cassel, R., Neidl, R., Selvi, B. R., Mohankrishna, D. V., Vedamurthy, B. M., Schneider, A., Bousiges, O., Mathis, C., Cassel, J.-C., Eswaramoorthy, M., Kundu, T. K., & Boutillier, A.-L. (2013). A novel activator of CBP/p300 acetyltransferases promotes neurogenesis and extends memory duration in adult mice. *Journal of Neuroscience, 33*(26), 10698–10712. https://doi.org/10.1523/JNEUROSCI.5772-12.2013

Chédin, F. (2011). The DNMT3 family of mammalian de novo DNA methyltransferases. *Progress in Molecular Biology and Translational Science, 101*, 255–285. https://doi.org/10.1016/B978-0-12-387685-0.00007-X

Chiao, J. Y. (2018). Developmental aspects in cultural neuroscience. *Developmental Review: DR, 50*(A), 77–89. https://doi.org/10.1016/j.dr.2018.06.005

Chivet, M., Hemming, F., Pernet-Gallay, K., Fraboulet, S., & Sadoul, R. (2012). Emerging role of neuronal exosomes in the central nervous system. *Frontiers in Physiology, 3*, 145. https://doi.org/10.3389/fphys.2012.00145

Choe, Y. (2014). Anti-Hebbian learning. In D. Jaeger & R. Jung (Eds.), *Encyclopedia of computational neuroscience* (pp. 1–4). Springer. https://doi.org/10.1007/978-1-4614-7320-6_675-1

Choi, D.-H., Choi, I.-A., & Lee, J. (2022). The role of DNA methylation in stroke recovery. *International Journal of Molecular Sciences, 23*(18), 10373. https://doi.org/10.3390/ijms231810373

Choudhury, S. (2010). Culturing the adolescent brain: What can neuroscience learn from anthropology? *Social Cognitive and Affective Neuroscience, 5*(2–3), 159–167. https://doi.org/10.1093/scan/nsp030

Choudhury, S., & Slaby, J. (2012). *Critical neuroscience: A handbook of the social and cultural contexts of neuroscience.* Wiley-Blackwell.

Chowdhury, S., Shepherd, J. D., Okuno, H., Lyford, G., Petralia, R. S., Plath, N., Kuhl, D., Huganir, R. L., & Worley, P. F. (2006). Arc/Arg3.1 interacts with the endocytic machinery to regulate AMPA receptor trafficking. *Neuron, 52*(3), 445–459. https://doi.org/10.1016/j.neuron.2006.08.033

Chugunow, L. (2022). Ivan P. Pavlov's dedication to science: Investigating the behavior and the human psyche. In J. Valsiner (Ed.), *One dog is enough. Ivan P. Pavlov's contribution to ideographic science* (pp. 1–45). Information Age Publishing.

Chung, L. (2015). A brief introduction to the transduction of neural activity into Fos signal. *Development & Reproduction, 19*(2), 61–67. https://doi.org/10.12717/DR.2015.19.2.061

Chwang, W. B., O'Riordan, K. J., Levenson, J. M., & Sweatt, J. D. (2006). ERK/MAPK regulates hippocampal histone phosphorylation following contextual fear conditioning. *Learning & Memory, 13*(3), 322–328. https://doi.org/10.1101/lm.152906

Cimino, G. (1999). Reticular theory versus neuron theory in the work of Camillo Golgi. *Physis; Rivista Internazionale Di Storia Della Scienza, 36*(2), 431–472.

Citri, A., & Malenka, R. C. (2008). Synaptic plasticity: Multiple forms, functions, and mechanisms. *Neuropsychopharmacology, 33*(1), Article 1. https://doi.org/10.1038/sj.npp.1301559

Clark, S. J., Lee, H. J., Smallwood, S. A., Kelsey, G., & Reik, W. (2016). Single-cell epigenomics: Powerful new methods for understanding gene regulation and cell identity. *Genome Biology, 17*(1), 72. https://doi.org/10.1186/s13059-016-0944-x

Clopath, C., Bonhoeffer, T., Hübener, M., & Rose, T. (2017). Variance and invariance of neuronal long-term representations. *Philosophical Transactions of the Royal Society B: Biological Sciences, 372*(1715), 20160161. https://doi.org/10.1098/rstb.2016.0161

Corner, M. A., & Ramakers, G. J. (1992). Spontaneous firing as an epigenetic factor in brain development—Physiological consequences of chronic tetrodotoxin and picrotoxin exposure on cultured rat neocortex neurons. *Brain Research. Developmental Brain Research, 65*(1), 57–64. https://doi.org/10.1016/0165-3806(92)90008-k

Cortés-Mendoza, J., Díaz de León-Guerrero, S., Pedraza-Alva, G., & Pérez-Martínez, L. (2013). Shaping synaptic plasticity: The role of activity-mediated epigenetic regulation on gene transcription. *International Journal of Developmental Neuroscience: The Official Journal of the International Society for Developmental Neuroscience, 31*(6), 359–369. https://doi.org/10.1016/j.ijdevneu.2013.04.003

Cowan, W., & Kandel, E. (2001). Prospects for neurology and psychiatry. *JAMA: The Journal of the American Medical Association, 285*, 594–600. https://doi.org/10.1001/jama.285.5.594

Cronbach, J. L. (1955). On the non-rational application of information measures in psychology. In H. Quastler (Ed.), *Information theory in psychology* (pp. 14–26). The Free Press.

Cullell, N., Soriano-Tárraga, C., Gallego-Fábrega, C., Cárcel-Márquez, J., Muiño, E., Llucià-Carol, L., Lledós, M., Esteller, M., de Moura, M. C., Montaner, J., Rosell, A., Delgado, P., Martí-Fábregas, J., Krupinski, J., Roquer, J., Jiménez-Conde, J., & Fernández-Cadenas, I. (2022). Altered methylation pattern in EXOC4 is associated with stroke outcome: An epigenome-wide association study. *Clinical Epigenetics, 14*(1), 124. https://doi.org/10.1186/s13148-022-01340-5

Dahmen, D., Layer, M., Deutz, L., Dąbrowska, P. A., Voges, N., von Papen, M., Brochier, T., Riehle, A., Diesmann, M., Grün, S., & Helias, M. (2022). Global organization of neuronal activity only requires unstructured local connectivity. *eLife, 11*, e68422. https://doi.org/10.7554/eLife.68422

Damasio, A. R. (2005). *Descartes' error: Emotion, reason and the human brain.* Penguin Books.

Damasio, A., Damasio, H., & Tranel, D. (2013). Persistence of feelings and sentience after bilateral damage of the insula. *Cerebral Cortex (New York, N.Y.: 1991), 23*(4), 833–846. https://doi.org/10.1093/cercor/bhs077

Danielsson, A., Nemes, S., Tisell, M., Lannering, B., Nordborg, C., Sabel, M., & Carén, H. (2015). MethPed: A DNA methylation classifier tool for the identification of pediatric brain tumor subtypes. *Clinical Epigenetics, 7*(1), 62. https://doi.org/10.1186/s13148-015-0103-3

Darmanis, S., Sloan, S. A., Zhang, Y., Enge, M., Caneda, C., Shuer, L. M., Hayden Gephart, M. G., Barres, B. A., & Quake, S. R. (2015). A survey of human brain transcriptome diversity at the single cell level. *Proceedings of the National Academy of Sciences, 112*(23), 7285–7290. https://doi.org/10.1073/pnas.1507125112

Das, A. (1997). Plasticity in adult sensory cortex: A review. *Network: Computation in Neural Systems, 8*(2), R33–R76. https://doi.org/10.1088/0954-898X_8_2_001

Dash, P. K., Hochner, B., & Kandel, E. R. (1990). Injection of the cAMP-responsive element into the nucleus of Aplysia sensory neurons blocks long-term facilitation. *Nature, 345*(6277), 718–721. https://doi.org/10.1038/345718a0

Day, J. J., & Sweatt, J. D. (2010). DNA methylation and memory formation. *Nature Neuroscience, 13*(11), 1319–1323. https://doi.org/10.1038/nn.2666

Day, J. J., & Sweatt, J. D. (2011a). Cognitive neuroepigenetics: A role for epigenetic mechanisms in learning and memory. *Neurobiology of Learning and Memory, 96*(1), 2–12. https://doi.org/10.1016/j.nlm.2010.12.008

Day, J. J., & Sweatt, J. D. (2011b). Epigenetic modifications in neurons are essential for formation and storage of behavioral memory. *Neuropsychopharmacology, 36*(1), 357–358. https://doi.org/10.1038/npp.2010.125

De Backer, J.-F., & Grunwald Kadow, I. C. (2022). A role for glia in cellular and systemic metabolism: Insights from the fly. *Current Opinion in Insect Science, 53*, 100947. https://doi.org/10.1016/j.cois.2022.100947

de Lima Camillo, L. P., Lapierre, L. R., & Singh, R. (2022). A pan-tissue DNA-methylation epigenetic clock based on deep learning. *Npj Aging, 8*(1), Article 1. https://doi.org/10.1038/s41514-022-00085-y

de Ruyter van Steveninck, R., Bialek, W., & Barlow, H. B. (1997). Real-time performance of a movement-sensitive neuron in the blowfly visual system: Coding and information transfer in short spike sequences. *Proceedings of the Royal Society of London. Series B. Biological Sciences, 234*(1277), 379–414. https://doi.org/10.1098/rspb.1988.0055

Deichmann, U. (2016). Epigenetics: The origins and evolution of a fashionable topic. *Developmental Biology, 416*(1), 249–254. https://doi.org/10.1016/j.ydbio.2016.06.005

Di Lullo, E., & Kriegstein, A. R. (2017). The use of brain organoids to investigate neural development and disease. *Nature Reviews Neuroscience, 18*(10), Article 10. https://doi.org/10.1038/nrn.2017.107

Dias, B. G., & Ressler, K. J. (2014). Parental olfactory experience influences behavior and neural structure in subsequent generations. *Nature Neuroscience, 17*(1), Article 1. https://doi.org/10.1038/nn.3594

Dieckmann, L., Cruceanu, C., Lahti-Pulkkinen, M., Lahti, J., Kvist, T., Laivuori, H., Sammallahti, S., Villa, P. M., Suomalainen-König, S., Rancourt, R. C., Plagemann, A., Henrich, W., Eriksson, J. G., Kajantie, E., Entringer, S., Braun, T., Räikkönen, K., Binder, E. B., & Czamara, D. (2022). Reliability of a novel approach for reference-based cell type estimation in human placental DNA methylation studies. *Cellular and Molecular Life Sciences, 79*(2), 115. https://doi.org/10.1007/s00018-021-04091-3

Dimitrov, A. G., Lazar, A. A., & Victor, J. D. (2011). Information theory in neuroscience. *Journal of Computational Neuroscience, 30*(1), 1–5. https://doi.org/10.1007/s10827-011-0314-3

Dineva, E., & Schöner, G. (2018). How infants' reaches reveal principles of sensorimotor decision making. *Connection Science, 30*(1), 53–80. https://doi.org/10.1080/09540091.2017.1405382

Doi, S., & Kumagai, S. (2005). Generation of very slow neuronal rhythms and chaos near the Hopf bifurcation in single neuron models. *Journal of Computational Neuroscience, 19*(3), 325–356. https://doi.org/10.1007/s10827-005-2895-1

Dunn, E. C., Soare, T. W., Zhu, Y., Simpkin, A. J., Suderman, M. J., Klengel, T., Smith, A. D. A. C., Ressler, K. J., & Relton, C. L. (2019). Sensitive periods for the effect of childhood adversity on DNA methylation: Results from a prospective, longitudinal study. *Biological Psychiatry, 85*(10), 838–849. https://doi.org/10.1016/j.biopsych.2018.12.023

Dunsmoor, J. E., Niv, Y., Daw, N., & Phelps, E. A. (2015). Rethinking extinction. *Neuron, 88*(1), 47–63. https://doi.org/10.1016/j.neuron.2015.09.028

Durchdewald, M., Angel, P., & Hess, J. (2009). The transcription factor Fos: A Janus-type regulator in health and disease. *Histology and Histopathology, 24*(11), 1451–1461. https://doi.org/10.14670/HH-24.1451

Dusart, I., & Flamant, F. (2012). Profound morphological and functional changes of rodent Purkinje cells between the first and the second postnatal weeks: A metamorphosis? *Frontiers in Neuroanatomy, 6*, 11. https://doi.org/10.3389/fnana.2012.00011

Edelstein, L., & Smythies, J. (2014). The role of epigenetic-related codes in neurocomputation: Dynamic hardware in the brain. *Philosophical Transactions of the Royal Society B: Biological Sciences, 369*(1652), 20130519. https://doi.org/10.1098/rstb.2013.0519

Elbert, T., Pantev, C., Wienbruch, C., Rockstroh, B., & Taub, E. (1995). Increased cortical representation of the fingers of the left hand in string players. *Science (New York, N.Y.), 270*(5234), 305–307. https://doi.org/10.1126/science.270.5234.305

Elsayed, M., & Magistretti, P. J. (2015). A new outlook on mental illnesses: Glial involvement beyond the glue. *Frontiers in Cellular Neuroscience, 9*, 468. https://www.frontiersin.org/articles/10.3389/fncel.2015.00468

Emery, E. C., Luiz, A. P., Sikandar, S., Magnúsdóttir, R., Dong, X., & Wood, J. N. (2016). In vivo characterization of distinct modality-specific subsets of somatosensory neurons using GCaMP. *Science Advances, 2*(11), e1600990. https://doi.org/10.1126/sciadv.1600990

Emsley, J. G., Mitchell, B. D., Kempermann, G., & Macklis, J. D. (2005). Adult neurogenesis and repair of the adult CNS with neural progenitors, precursors, and stem cells. *Progress in Neurobiology, 75*(5), 321–341. https://doi.org/10.1016/j.pneurobio.2005.04.002

Enander, J. M. D., & Jörntell, H. (2019). Somatosensory cortical neurons decode tactile input patterns and location from both dominant and non-dominant digits. *Cell Reports, 26*(13), 3551–3560.e4. https://doi.org/10.1016/j.celrep.2019.02.099

Endres, M., Fan, G., Meisel, A., Dirnagl, U., & Jaenisch, R. (2001). Effects of cerebral ischemia in mice lacking DNA methyltransferase 1 in post-mitotic neurons. *Neuroreport, 12*(17), 3763–3766. https://doi.org/10.1097/00001756-200112040-00032

Engel, A. K., Fries, P., König, P., Brecht, M., & Singer, W. (1999). Temporal binding, binocular rivalry, and consciousness. *Consciousness and Cognition, 8*(2), 128–151. https://doi.org/10.1006/ccog.1999.0389

Eppe, M., Wermter, S., Hafner, V. V., & Nagai, Y. (2021). Developmental robotics and its role towards artificial general intelligence. *KI – Künstliche Intelligenz, 35*(1), 5–7. https://doi.org/10.1007/s13218-021-00706-w

Erikson, E. H. (1959). *Identity and the life cycle: Selected papers.* International Universities Press.

Erikson, E. H. (1968). *Identity: Youth and crisis.* Norton & Co.

Erzurumlu, R. S., & Gaspar, P. (2020). How the barrel cortex became a working model for developmental plasticity: A historical perspective. *Journal of Neuroscience, 40*(34), 6460–6473. https://doi.org/10.1523/JNEUROSCI.0582-20.2020

Espinosa, J. S., & Stryker, M. P. (2012). Development and plasticity of the primary visual cortex. *Neuron, 75*(2), 230–249. https://doi.org/10.1016/j.neuron.2012.06.009

Faber, D. S., & Pereda, A. E. (2018). Two forms of electrical transmission between neurons. *Frontiers in Molecular Neuroscience, 11*, 427. https://doi.org/10.3389/fnmol.2018.00427

Farhy-Tselnicker, I., & Allen, N. J. (2018). Astrocytes, neurons, synapses: A tripartite view on cortical circuit development. *Neural Development, 13*(1), 7. https://doi.org/10.1186/s13064-018-0104-y

Feng, J., Chang, H., Li, E., & Fan, G. (2005). Dynamic expression of de novo DNA methyltransferases Dnmt3a and Dnmt3b in the central nervous system. *Journal of Neuroscience Research, 79*(6), 734–746. https://doi.org/10.1002/jnr.20404

Feng, J., Zhou, Y., Campbell, S. L., Le, T., Li, E., Sweatt, J. D., Silva, A. J., & Fan, G. (2010). Dnmt1 and Dnmt3a maintain DNA methylation and regulate synaptic function in adult forebrain neurons. *Nature Neuroscience, 13*(4), Article 4. https://doi.org/10.1038/nn.2514

Feuillet, L., Dufour, H., & Pelletier, J. (2007). Brain of a white-collar worker. *The Lancet, 370*(9583), 262. https://doi.org/10.1016/S0140-6736(07)61127-1

Fiedler, K., Kliegl, R., Lindenberger, U., Mausfeld, R., Mummendey, A., & Prinz, W. (2005). Psychologie im 21. Jahrhundert: Führende deutsche Psychologen über Lage und Zukunft ihres Fachs und die Rolle der psychologischen Grundlagenforschung. *Gehirn & Geist, 7–8*, 56–60.

Fields, R. D., & Stevens-Graham, B. (2002). New insights into neuron-glia communication. *Science (New York, N.Y.), 298*(5593), 556–562. https://doi.org/10.1126/science.298.5593.556

Fields, R. D., Woo, D. H., & Basser, P. J. (2015). Glial regulation of the neuronal connectome through local and long-distant communication. *Neuron, 86*(2), 374–386. https://doi.org/10.1016/j.neuron.2015.01.014

Flament, S. (2016). Sex reversal in amphibians. *Sexual Development, 10*(5–6), 267–278. https://doi.org/10.1159/000448797

Flavahan, W. A. (2020). Epigenetic plasticity, selection, and tumorigenesis. *Biochemical Society Transactions, 48*(4), 1609–1621. https://doi.org/10.1042/BST20191215

Flavahan, W. A., Gaskell, E., & Bernstein, B. E. (2017). Epigenetic plasticity and the hallmarks of cancer. *Science, 357*(6348), eaal2380. https://doi.org/10.1126/science.aal2380

Fodstad, H. (2001). The neuron theory. *Stereotactic and Functional Neurosurgery, 77*(1–4), 20–24. https://doi.org/10.1159/000064596

Fogassi, L., Ferrari, P., Gesierich, B., Rozzi, S., Chersi, F., & Rizzolatti, G. (2005). Parietal lobe: From action organization to intention understanding. *Science (New York, N.Y.), 308*(5722), 662–667. https://doi.org/10.1126/science.1106138

Fraga, M. F., Ballestar, E., Paz, M. F., Ropero, S., Setien, F., Ballestar, M. L., Heine-Suñer, D., Cigudosa, J. C., Urioste, M., Benitez, J., Boix-Chornet, M., Sanchez-Aguilera, A., Ling, C., Carlsson, E., Poulsen, P., Vaag, A., Stephan, Z., Spector, T. D., Wu, Y.-Z., et al. (2005). Epigenetic differences arise during the lifetime of monozygotic twins. *Proceedings of the National Academy of Sciences of the United States of America, 102*(30), 10604. https://doi.org/10.1073/pnas.0500398102

Franklin, T. B., & Mansuy, I. M. (2010). Epigenetic inheritance in mammals. *Neurobiology of Disease, 39*(1), 61–65. https://doi.org/10.1016/j.nbd.2009.11.012

Franks, K. M., & Isaacson, J. S. (2005). Synapse-specific downregulation of NMDA receptors by early experience: A critical period for plasticity of sensory input to olfactory cortex. *Neuron, 47*(1), 101–114. https://doi.org/10.1016/j.neuron.2005.05.024

Fransquet, P. D., Wrigglesworth, J., Woods, R. L., Ernst, M. E., & Ryan, J. (2019). The epigenetic clock as a predictor of disease and mortality risk: A systematic review and meta-analysis. *Clinical Epigenetics, 11*(1), 62. https://doi.org/10.1186/s13148-019-0656-7

Freud, S. (1895). A project for a scientific psychology. In *Standard edition* (Vol. I, pp. 283–397). Hogarth.

Fries, P. (2015). Rhythms for cognition: Communication through coherence. *Neuron, 88*(1), 220–235. https://doi.org/10.1016/j.neuron.2015.09.034

Friston, K. J. (1997). Another neural code? *NeuroImage, 5*(3), 213–220. https://doi.org/10.1006/nimg.1997.0260

Friston, K. J., Stephan, K. E., Montague, R., & Dolan, R. J. (2014). Computational psychiatry: The brain as a phantastic organ. *The Lancet. Psychiatry, 1*(2), 148–158. https://doi.org/10.1016/S2215-0366(14)70275-5

Frühbeis, C., Fröhlich, D., Kuo, W. P., & Krämer-Albers, E.-M. (2013). Extracellular vesicles as mediators of neuron-glia communication. *Frontiers in Cellular Neuroscience, 7*, 182. https://doi.org/10.3389/fncel.2013.00182

Fuchs, T. (2018). *Ecology of the brain: The phenomenology and biology of the embodied mind* (1st ed.). Oxford University Press.

Fuchs, T. (2020). The circularity of the embodied mind. *Frontiers in Psychology, 11*, 1707. https://doi.org/10.3389/fpsyg.2020.01707

Fuchs, E. C., Neitz, A., Pinna, R., Melzer, S., Caputi, A., & Monyer, H. (2016). Local and distant input controlling excitation in layer II of the medial entorhinal cortex. *Neuron, 89*(1), 194–208. https://doi.org/10.1016/j.neuron.2015.11.029

Galkin, F., Mamoshina, P., Kochetov, K., Sidorenko, D., & Zhavoronkov, A. (2021). DeepMAge: A methylation aging clock developed with deep learning. *Aging and Disease, 12*(5), 1252–1262. https://doi.org/10.14336/AD.2020.1202

Gallese, V. (2013). Mirror neurons, embodied simulation and a second-person approach to mind-reading. *Cortex, 49*(10), 2954–2956. https://doi.org/10.1016/j.cortex.2013.09.008

Georgopoulos, A. P., Schwartz, A. B., & Kettner, R. E. (1986). Neuronal population coding of movement direction. *Science, 233*(4771), 1416–1419. https://doi.org/10.1126/science.3749885

Gerhard, F., Haslinger, R., & Pipa, G. (2011). Applying the multivariate time-rescaling theorem to neural population models. *Neural Computation, 23*(6), 1452–1483. https://doi.org/10.1162/NECO_a_00126

Gerstein, M. B., Bruce, C., Rozowsky, J. S., Zheng, D., Du, J., Korbel, J. O., Emanuelsson, O., Zhang, Z. D., Weissman, S., & Snyder, M. (2007). What is a gene, post-ENCODE? History and updated definition. *Genome Research, 17*(6), 669–681. https://doi.org/10.1101/gr.6339607

Gerstner, W., Kreiter, A. K., Markram, H., & Herz, A. V. M. (1997). Neural codes: Firing rates and beyond. *Proceedings of the National Academy of Sciences, 94*(24), 12740–12741. https://doi.org/10.1073/pnas.94.24.12740

Geyer, C. (Ed.). (2004). *Hirnforschung und Willensfreiheit: Zur Deutung der neuesten Experimente* (9. Auflage, Originalausgabe). Suhrkamp.

Ghahramani, N. M., Ngun, T. C., Chen, P.-Y., Tian, Y., Krishnan, S., Muir, S., Rubbi, L., Arnold, A. P., de Vries, G. J., Forger, N. G., Pellegrini, M., & Vilain, E. (2014). The effects of perinatal testosterone exposure on the DNA methylome of the mouse brain are late-emerging. *Biology of Sex Differences, 5*(1), 8. https://doi.org/10.1186/2042-6410-5-8

Gilbert, S. F. (2001). Ecological developmental biology: Developmental biology meets the real world. *Developmental Biology, 233*(1), 1–12. https://doi.org/10.1006/dbio.2001.0210

Gilbert, S. F., & Barresi, M. J. F. (2020). *Developmental biology* (12th ed.). Oxford University Press.

Gilbert, S. F., Bosch, T. C. G., & Ledón-Rettig, C. (2015). Eco-Evo-Devo: Developmental symbiosis and developmental plasticity as evolutionary agents. *Nature Reviews. Genetics, 16*(10), 611–622. https://doi.org/10.1038/nrg3982

Gluckman, P. D., Hanson, M. A., Spencer, H. G., & Bateson, P. (2005). Environmental influences during development and their later consequences for health and disease: Implications for the interpretation of empirical studies. *Proceedings. Biological sciences, 272*(1564), 671–677. https://doi.org/10.1098/rspb.2004.3001

Goldin, M. A., Harrell, E. R., Estebanez, L., & Shulz, D. E. (2018). Rich spatio-temporal stimulus dynamics unveil sensory specialization in cortical area S2. *Nature Communications, 9*(1), Article 1. https://doi.org/10.1038/s41467-018-06585-4

Gordon, E. (2003). Integrative neuroscience. *Neuropsychopharmacology, 28*(1), Article 1. https://doi.org/10.1038/sj.npp.1300136

Gorkin, D. U., Qiu, Y., Hu, M., Fletez-Brant, K., Liu, T., Schmitt, A. D., Noor, A., Chiou, J., Gaulton, K. J., Sebat, J., Li, Y., Hansen, K. D., & Ren, B. (2019). Common DNA sequence variation influences 3-dimensional conformation of the human genome. *Genome Biology, 20*(1), 255. https://doi.org/10.1186/s13059-019-1855-4

Goto, K., Numata, M., Komura, J. I., Ono, T., Bestor, T. H., & Kondo, H. (1994). Expression of DNA methyltransferase gene in mature and immature neurons as well as proliferating cells in mice. *Differentiation; Research in Biological Diversity, 56*(1–2), 39–44. https://doi.org/10.1046/j.1432-0436.1994.56120039.x

Gottlieb, G. (2007). Probabilistic epigenesis. *Developmental Science, 10*(1), 1–11. https://doi.org/10.1111/j.1467-7687.2007.00556.x

Goyal, D., Limesand, S. W., & Goyal, R. (2019). Epigenetic responses and the developmental origins of health and disease. *Journal of Endocrinology, 242*(1), T105–T119. https://doi.org/10.1530/JOE-19-0009

Gräff, J., & Mansuy, I. M. (2008). Epigenetic codes in cognition and behaviour. *Behavioural Brain Research, 192*(1), 70–87. https://doi.org/10.1016/j.bbr.2008.01.021

Gräff, J., & Tsai, L.-H. (2013). Histone acetylation: Molecular mnemonics on the chromatin. *Nature Reviews. Neuroscience, 14*(2), 97–111. https://doi.org/10.1038/nrn3427

Gray, C. M., & Singer, W. (1989). Stimulus-specific neuronal oscillations in orientation columns of cat visual cortex. *Proceedings of the National Academy of Sciences, 86*(5), 1698–1702. https://doi.org/10.1073/pnas.86.5.1698

Greenough, W., Black, J., & Wallace, C. (1987). Experience and brain development. *Child Development, 58*(3), 539–559. https://pubmed.ncbi.nlm.nih.gov/3038480/

Grefkes, C., & Fink, G. R. (2020). Recovery from stroke: Current concepts and future perspectives. *Neurological Research and Practice, 2*(1), 17. https://doi.org/10.1186/s42466-020-00060-6

Griffiths, P. E., & Hochman, A. (2015). Developmental systems theory. In *ELS* (pp. 1–7). Wiley. https://doi.org/10.1002/9780470015902.a0003452.pub2

Grodstein, F., Lemos, B., Yu, L., Iatrou, A., De Jager, P. L., & Bennett, D. A. (2021). Characteristics of epigenetic clocks across blood and brain tissue in older women and men. *Frontiers in Neuroscience, 14*, 555307. https://www.frontiersin.org/articles/10.3389/fnins.2020.555307

Grover, S., Nguyen, J. A., & Reinhart, R. M. G. (2021). Synchronizing brain rhythms to improve cognition. *Annual Review of Medicine, 72*, 29–43. https://doi.org/10.1146/annurev-med-060619-022857

Guan, Z., Giustetto, M., Lomvardas, S., Kim, J.-H., Miniaci, M. C., Schwartz, J. H., Thanos, D., & Kandel, E. R. (2002). Integration of long-term-memory-related synaptic plasticity involves bidirectional regulation of gene expression and chromatin structure. *Cell, 111*(4), 483–493. https://doi.org/10.1016/s0092-8674(02)01074-7

Guerra-Carrillo, B., Mackey, A. P., & Bunge, S. A. (2014). Resting-state fMRI: A window into human brain plasticity. *The Neuroscientist: A Review Journal Bringing Neurobiology, Neurology and Psychiatry, 20*(5), 522–533. https://doi.org/10.1177/1073858414524442

Guic, E., Carrasco, X., Rodríguez, E., Robles, I., & Merzenich, M. M. (2008). Plasticity in primary somatosensory cortex resulting from environmentally enriched stimulation and sensory discrimination training. *Biological Research, 41*(4), 425–437. https://doi.org/10.4067/S0716-97602008000400008

Guillery, R. W. (2005). Observations of synaptic structures: Origins of the neuron doctrine and its current status. *Philosophical Transactions of the Royal Society of London. Series B, Biological Sciences, 360*(1458), 1281–1307. https://doi.org/10.1098/rstb.2003.1459

Guillery, R. W. (2007). Relating the neuron doctrine to the cell theory. Should contemporary knowledge change our view of the neuron doctrine? *Brain Research Reviews, 55*(2), 411–421. https://doi.org/10.1016/j.brainresrev.2007.01.005

Haggard, P., & Libet, B. (2001). Conscious intention and brain activity. *Journal of Consciousness Studies, 8*(11), 47–63.

Hagmann, P. (2005). *From diffusion MRI to brain connectomics* (Thèse no 3230). EPFL, Lausanne. https://doi.org/10.5075/epfl-thesis-3230

Hagmann, P., Cammoun, L., Gigandet, X., Gerhard, S., Grant, P. E., Wedeen, V., Meuli, R., Thiran, J.-P., Honey, C. J., & Sporns, O. (2010). MR connectomics: Principles and challenges. *Journal of Neuroscience Methods, 194*(1), 34–45. https://doi.org/10.1016/j.jneumeth.2010.01.014

Hales, C. N., & Barker, D. J. (1992). Type 2 (non-insulin-dependent) diabetes mellitus: The thrifty phenotype hypothesis. *Diabetologia, 35*(7), 595–601. https://doi.org/10.1007/BF00400248

Hannum, G., Guinney, J., Zhao, L., Zhang, L., Hughes, G., Sadda, S., Klotzle, B., Bibikova, M., Fan, J.-B., Gao, Y., Deconde, R., Chen, M., Rajapakse, I., Friend, S., Ideker, T., & Zhang, K. (2013). Genome-wide methylation profiles reveal quantitative views of human aging rates. *Molecular Cell, 49*(2), 359–367. https://doi.org/10.1016/j.molcel.2012.10.016

Hansel, C., Linden, D. J., & D'Angelo, E. (2001). Beyond parallel fiber LTD: The diversity of synaptic and non-synaptic plasticity in the cerebellum. *Nature Neuroscience, 4*(5), 467–475. https://doi.org/10.1038/87419

Hanson, M. A., & Gluckman, P. D. (2014). Early developmental conditioning of later health and disease: Physiology or pathophysiology? *Physiological Reviews, 94*(4), 1027–1076. https://doi.org/10.1152/physrev.00029.2013

Hawrylycz, M. J., Lein, E. S., Guillozet-Bongaarts, A. L., Shen, E. H., Ng, L., Miller, J. A., van de Lagemaat, L. N., Smith, K. A., Ebbert, A., Riley, Z. L., Abajian, C., Beckmann, C. F., Bernard, A., Bertagnolli, D., Boe, A. F., Cartagena, P. M., Chakravarty, M. M., Chapin, M., Chong, J., et al. (2012). An anatomically comprehensive atlas of the adult human brain transcriptome. *Nature, 489*(7416), 391–399. https://doi.org/10.1038/nature11405

Hayashi-Takanaka, Y., Kina, Y., Nakamura, F., Becking, L. E., Nakao, Y., Nagase, T., Nozaki, N., & Kimura, H. (2020). Histone modification dynamics as revealed by multicolor immunofluorescence-based single-cell analysis. *Journal of Cell Science, 133*(14), jcs243444. https://doi.org/10.1242/jcs.243444

Heard, E., & Martienssen, R. A. (2014). Transgenerational epigenetic inheritance: Myths and mechanisms. *Cell, 157*(1), 95–109. https://doi.org/10.1016/j.cell.2014.02.045

Hebb, D. O. (1949). *The organization of behavior. A neuropsychological theory*. Wiley.

Heim, C., & Binder, E. B. (2012). Current research trends in early life stress and depression. *Experimental Neurology, 233*(1), 102–111. https://doi.org/10.1016/j.expneurol.2011.10.032

Heiss, J. A., Brennan, K. J., Baccarelli, A. A., Téllez-Rojo, M. M., Estrada-Gutiérrez, G., Wright, R. O., & Just, A. C. (2019). Battle of epigenetic proportions: Comparing Illumina's EPIC methylation microarrays and TruSeq targeted bisulfite sequencing. *Epigenetics, 15*(1–2), 174–182. https://doi.org/10.1080/15592294.2019.1656159

Hempel, C. M., Sugino, K., & Nelson, S. B. (2007). A manual method for the purification of fluorescently labeled neurons from the mammalian brain. *Nature Protocols, 2*(11), 2924–2929. https://doi.org/10.1038/nprot.2007.416

Herrero-Navarro, Á., Puche-Aroca, L., Moreno-Juan, V., Sempere-Ferràndez, A., Espinosa, A., Susín, R., Torres-Masjoan, L., Leyva-Díaz, E., Karow, M., Figueres-Oñate, M., López-Mascaraque, L., López-Atalaya, J. P., Berninger, B., & López-Bendito, G. (2021). Astrocytes and neurons share region-specific transcriptional signatures that confer regional identity to neuronal reprogramming. *Science Advances, 7*(15), eabe8978. https://doi.org/10.1126/sciadv.abe8978

Hines, M. L., & Carnevale, N. T. (2001). Neuron. *The Neuroscientist: A Review Journal Bringing Neurobiology, Neurology and Psychiatry, 7*(2), 123–135. https://doi.org/10.1177/107385840100700207

His, W. (2017). *Die Formentwickelung des menschlichen Vorderhirns: Vom Ende des ersten bis zum beginn des dritten Monats/Wilhelm His* (Nachdruck der Ausgabe von 1889). Hansebooks GmbH. http://nbn-resolving.de/urn:nbn:de:101:1-2018122721520047540280

Hoffman, D. J., Reynolds, R. M., & Hardy, D. B. (2017). Developmental origins of health and disease: Current knowledge and potential mechanisms. *Nutrition Reviews, 75*(12), 951–970. https://doi.org/10.1093/nutrit/nux053

Hoffmeyer, J. (2008). *Biosemiotics: An examination into the signs of life and the life of signs*. University of Scranton Press.

Hooks, B. M., & Chen, C. (2007). Critical periods in the visual system: Changing views for a model of experience-dependent plasticity. *Neuron, 56*(2), 312–326. https://doi.org/10.1016/j.neuron.2007.10.003

Hormuzdi, S. G., Filippov, M. A., Mitropoulou, G., Monyer, H., & Bruzzone, R. (2004). Electrical synapses: A dynamic signaling system that shapes the activity of neuronal networks. *Biochimica et Biophysica Acta (BBA) – Biomembranes, 1662*(1), 113–137. https://doi.org/10.1016/j.bbamem.2003.10.023

Horsthemke, B. (2018). A critical view on transgenerational epigenetic inheritance in humans. *Nature Communications, 9*(1), Article 1. https://doi.org/10.1038/s41467-018-05445-5

Horsthemke, B. (2022). A critical appraisal of clinical epigenetics. *Clinical Epigenetics, 14*(1), 95. https://doi.org/10.1186/s13148-022-01315-6

Horvath, S. (2013). DNA methylation age of human tissues and cell types. *Genome Biology, 14*(10), R115. https://doi.org/10.1186/gb-2013-14-10-r115

Hsu, C.-N., & Tain, Y.-L. (2021). Animal models for DOHaD research: Focus on hypertension of developmental origins. *Biomedicine, 9*(6), 623. https://doi.org/10.3390/biomedicines9060623

Huang, Y., Yan, J., Hou, J., Fu, X., Li, L., & Hou, Y. (2015). Developing a DNA methylation assay for human age prediction in blood and bloodstain. *Forensic Science International. Genetics, 17*, 129–136. https://doi.org/10.1016/j.fsigen.2015.05.007

Huang, R.-C., Lillycrop, K. A., Beilin, L. J., Godfrey, K. M., Anderson, D., Mori, T. A., Rauschert, S., Craig, J. M., Oddy, W. H., Ayonrinde, O. T., Pennell, C. E., Holbrook, J. D., & Melton, P. E. (2019). Epigenetic age acceleration in adolescence associates with BMI, inflammation, and risk score for middle age cardiovascular disease. *The Journal of Clinical Endocrinology and Metabolism, 104*(7), 3012–3024. https://doi.org/10.1210/jc.2018-02076

Hubel, D. H., & Wiesel, T. N. (1963). Receptive fields of cells in striate cortex of very young, visually inexperienced kittens. *Journal of Neurophysiology, 26*, 994–1002.

Ibarra-Lecue, I., Haegens, S., & Harris, A. Z. (2022). Breaking down a rhythm: Dissecting the mechanisms underlying task-related neural oscillations. *Frontiers in Neural Circuits, 16*, 846905. https://www.frontiersin.org/articles/10.3389/fncir.2022.846905

Isles, A. R. (2015). Neural and behavioral epigenetics; what it is, and what is hype. *Genes, Brain, and Behavior, 14*(1), 64–72. https://doi.org/10.1111/gbb.12184

Isles, A. R., & Wilkinson, L. S. (2008). Epigenetics: What is it and why is it important to mental disease? *British Medical Bulletin, 85*(1), 35–45. https://doi.org/10.1093/bmb/ldn004

Ittel, A., & Kretschmer, T. (2007). Historical roots of developmental science. *International Journal of Developmental Science, 1*(1), 23–32. https://doi.org/10.3233/DEV-2007-1103

Iyengar, S. (2003). The analysis of multiple neural spike trains. In *Advances on methodological and applied aspects of probability and statistics*. CRC Press.

Jablonka, E., & Lamb, M. J. (2005). *Evolution in four dimensions: Genetic, epigenetic, behavioral, and symbolic variation in the history of life* (pp. x, 462). MIT Press.

Jaffe, A. E., & Irizarry, R. A. (2014). Accounting for cellular heterogeneity is critical in epigenome-wide association studies. *Genome Biology, 15*(2), R31. https://doi.org/10.1186/gb-2014-15-2-r31

James, W. (1890). *The principles of psychology* (Vol. 1). Henry Holt and Company.

Jang, H. S., Shin, W. J., Lee, J. E., & Do, J. T. (2017). CpG and non-CpG methylation in epigenetic gene regulation and brain function. *Genes, 8*(6), 148. https://doi.org/10.3390/genes8060148

Jarosiewicz, B., Chase, S. M., Fraser, G. W., Velliste, M., Kass, R. E., & Schwartz, A. B. (2008). Functional network reorganization during learning in a brain-computer interface paradigm. *Proceedings of the National Academy of Sciences, 105*(49), 19486–19491. https://doi.org/10.1073/pnas.0808113105

Jenkinson, G., Pujadas, E., Goutsias, J., & Feinberg, A. P. (2017). Potential energy landscapes identify the information-theoretic nature of the epigenome. *Nature Genetics, 49*(5), Article 5. https://doi.org/10.1038/ng.3811

Jenkinson, G., Abante, J., Feinberg, A. P., & Goutsias, J. (2018). An information-theoretic approach to the modeling and analysis of whole-genome bisulfite sequencing data. *BMC Bioinformatics, 19*(1), 87. https://doi.org/10.1186/s12859-018-2086-5

Jenkinson, G., Abante, J., Koldobskiy, M. A., Feinberg, A. P., & Goutsias, J. (2019). Ranking genomic features using an information-theoretic measure of epigenetic discordance. *BMC Bioinformatics, 20*(1), 175. https://doi.org/10.1186/s12859-019-2777-6

Johnson, M. H. (2000). Functional brain development in infants: Elements of an interactive specialization framework. *Child Development, 71*(1), 75–81.

Johnson, M. H. (2011). Interactive specialization: A domain-general framework for human functional brain development? *Developmental Cognitive Neuroscience, 1*(1), 7–21. https://doi.org/10.1016/j.dcn.2010.07.003

Johnson, M. H. (2020). Chapter 13—Theories in developmental cognitive neuroscience. In J. Rubenstein, P. Rakic, B. Chen, & K. Y. Kwan (Eds.), *Neural circuit and cognitive development* (2nd ed., pp. 273–288). Academic Press. https://doi.org/10.1016/B978-0-12-814411-4.00013-5

Johnson, M. H., & de Haan, M. (2015). *Developmental cognitive neuroscience: An introduction*. Wiley.

Johnson, C., Kretsge, L. N., Yen, W. W., Sriram, B., O'Connor, A., Liu, R. S., Jimenez, J. C., Phadke, R. A., Wingfield, K. K., Yeung, C., Jinadasa, T. J., Nguyen, T. P. H., Cho, E. S., Fuchs, E., Spevack, E. D., Velasco, B. E., Hausmann, F. S., Fournier, L. A., Brack, A., et al. (2022). Highly unstable heterogeneous representations in VIP interneurons of the anterior cingulate cortex. *Molecular Psychiatry, 27*(5), 2602–2618. https://doi.org/10.1038/s41380-022-01485-y

Johnstone, S. E., Gladyshev, V. N., Aryee, M. J., & Bernstein, B. E. (2022). Epigenetic clocks, aging, and cancer. *Science, 378*(6626), 1276–1277. https://doi.org/10.1126/science.abn4009

Jones, E. G. (1994). The neuron doctrine 1891. *Journal of the History of the Neurosciences, 3*(1), 3–20. https://doi.org/10.1080/09647049409525584

Jones, P. A., & Baylin, S. B. (2007). The epigenomics of cancer. *Cell, 128*(4), 683–692. https://doi.org/10.1016/j.cell.2007.01.029

Jost, J. (2020). Biological information. *Theory in Biosciences = Theorie in Den Biowissenschaften, 139*(4), 361–370. https://doi.org/10.1007/s12064-020-00327-1

Jost, J. (2021). Information theory and consciousness. *Frontiers in Applied Mathematics and Statistics, 7*, 641239. https://www.frontiersin.org/articles/10.3389/fams.2021.641239

Jost, J., & Scherrer, K. (2014). Information theory, gene expression, and combinatorial regulation: A quantitative analysis. *Theory in Biosciences = Theorie in Den Biowissenschaften, 133*(1), 1–21. https://doi.org/10.1007/s12064-013-0182-7

Jurjuţ, O. F., Nikolić, D., Singer, W., Yu, S., Havenith, M. N., & Mureşan, R. C. (2011). Timescales of multineuronal activity patterns reflect temporal structure of visual stimuli. *PLoS One, 6*(2), e16758. https://doi.org/10.1371/journal.pone.0016758

Kadoshima, T., Sakaguchi, H., Nakano, T., Soen, M., Ando, S., Eiraku, M., & Sasai, Y. (2013). Self-organization of axial polarity, inside-out layer pattern, and species-specific progenitor dynamics in human ES cell-derived neocortex. *Proceedings of the National Academy of Sciences of the United States of America, 110*(50), 20284–20289. https://doi.org/10.1073/pnas.1315710110

Kahali, S., Raichle, M. E., & Yablonskiy, D. A. (2021). The role of the human brain neuron–glia–synapse composition in forming resting-state functional connectivity networks. *Brain Sciences, 11*(12), 1565. https://doi.org/10.3390/brainsci11121565

Kaiser, M., & Cromby, J. (2014). Neuroscience. In T. Teo (Ed.), *Encyclopedia of critical psychology* (pp. 1243–1248). Springer. https://doi.org/10.1007/978-1-4614-5583-7_200

Kandel, E. R. (2000). *The molecular biology of memory storage: A dialogue between genes and synapses. Nobel lecture.* https://www.nobelprize.org/uploads/2018/06/kandel-lecture.pdf

Karemaker, I. D., & Vermeulen, M. (2018). Single-cell DNA methylation profiling: Technologies and biological applications. *Trends in Biotechnology, 36*(9), 952–965. https://doi.org/10.1016/j.tibtech.2018.04.002

Kasabov, N. K. (2018). *Time-space, spiking neural networks and brain-inspired artificial intelligence*. Springer.

Kato, Y., Kaneda, M., Hata, K., Kumaki, K., Hisano, M., Kohara, Y., Okano, M., Li, E., Nozaki, M., & Sasaki, H. (2007). Role of the Dnmt3 family in de novo methylation of imprinted and repetitive sequences during male germ cell development in the mouse. *Human Molecular Genetics, 16*(19), 2272–2280. https://doi.org/10.1093/hmg/ddm179

Keck, T., Toyoizumi, T., Chen, L., Doiron, B., Feldman, D. E., Fox, K., Gerstner, W., Haydon, P. G., Hübener, M., Lee, H.-K., Lisman, J. E., Rose, T., Sengpiel, F., Stellwagen, D., Stryker, M. P., Turrigiano, G. G., & van Rossum, M. C. (2017). Integrating Hebbian and homeostatic plasticity: The current state of the field and future research directions. *Philosophical Transactions of the Royal Society of London. Series B, Biological Sciences, 372*(1715), 20160158. https://doi. org/10.1098/rstb.2016.0158

Keller, L., & Ross, K. G. (1993). Phenotypic plasticity and "cultural transmission" of alternative social organizations in the fire ant Solenopsis invicta. *Behavioral Ecology and Sociobiology, 33*(2), 121–129. https://doi.org/10.1007/BF00171663

Kemenes, I., Straub, V. A., Nikitin, E. S., Staras, K., O'Shea, M., Kemenes, G., & Benjamin, P. R. (2006). Role of delayed nonsynaptic neuronal plasticity in long-term associative memory. *Current Biology: CB, 16*(13), 1269–1279. https://doi.org/10.1016/j.cub.2006.05.049

Kempermann, G., Gage, F. H., Aigner, L., Song, H., Curtis, M. A., Thuret, S., Kuhn, H. G., Jessberger, S., Frankland, P. W., Cameron, H. A., Gould, E., Hen, R., Abrous, D. N., Toni, N., Schinder, A. F., Zhao, X., Lucassen, P. J., & Frisén, J. (2018). Human adult neurogenesis: Evidence and remaining questions. *Cell Stem Cell, 23*(1), 25–30. https://doi.org/10.1016/j. stem.2018.04.004

Kenet, T., Arieli, A., Tsodyks, M., & Grinvald, A. (2006). Are single cortical neurons soloists or are they obedient members of a huge orchestra? In J. L. van Hemmen & T. J. Sejnowski (Eds.), *23 Problems in systems neuroscience* (pp. 160–181). Oxford University Press. https:// doi.org/10.1093/acprof:oso/9780195148220.003.0009

Khodadadi, E., Fahmideh, L., Khodadadi, E., Dao, S., Yousefi, M., Taghizadeh, S., Asgharzadeh, M., Yousefi, B., & Kafil, H. S. (2021). Current advances in DNA methylation analysis methods. *BioMed Research International, 2021*, e8827516. https://doi.org/10.1155/2021/8827516

Kim, H. S., & Sasaki, J. Y. (2014). Cultural neuroscience: Biology of the mind in cultural contexts. *Annual Review of Psychology, 65*, 487–514. https://doi.org/10.1146/ annurev-psych-010213-115040

Kirkpatrick, J., Pascanu, R., Rabinowitz, N., Veness, J., Desjardins, G., Rusu, A. A., Milan, K., Quan, J., Ramalho, T., Grabska-Barwinska, A., Hassabis, D., Clopath, C., Kumaran, D., & Hadsell, R. (2017). Overcoming catastrophic forgetting in neural networks. *Proceedings of the National Academy of Sciences, 114*(13), 3521–3526. https://doi.org/10.1073/pnas.1611835114

Kirmayer, L. J. (2011). The future of critical neuroscience. In *Critical neuroscience* (pp. 367–383). Wiley. https://doi.org/10.1002/9781444343359.ch18

Kirmayer, L. J., & Gómez-Carrillo, A. (2019). Agency, embodiment and enactment in psychosomatic theory and practice. *Medical Humanities, 45*(2), 169–182. https://doi.org/10.1136/ medhum-2018-011618

Kitayama, S., & Park, J. (2010). Cultural neuroscience of the self: Understanding the social grounding of the brain. *Social Cognitive and Affective Neuroscience, 5*(2–3), 111–129. https:// doi.org/10.1093/scan/nsq052

Kitayama, S., & Park, J. (2014). Error-related brain activity reveals self-centric motivation: Culture matters. *Journal of Experimental Psychology. General, 143*(1), 62–70. https://doi.org/10.1037/ a0031696

Kitayama, S., & Tompson, S. (2010). Envisioning the future of cultural neuroscience. *Asian Journal of Social Psychology, 13*(2), 92–101. https://doi.org/10.1111/j.1467-839X.2010.01304.x

Klingenberg, C. P. (2019). Phenotypic plasticity, developmental instability, and robustness: The concepts and how they are connected. *Frontiers in Ecology and Evolution, 7*, 56. https://www. frontiersin.org/articles/10.3389/fevo.2019.00056

Knight, A. K., Craig, J. M., Theda, C., Bækvad-Hansen, M., Bybjerg-Grauholm, J., Hansen, C. S., Hollegaard, M. V., Hougaard, D. M., Mortensen, P. B., Weinsheimer, S. M., Werge, T. M., Brennan, P. A., Cubells, J. F., Newport, D. J., Stowe, Z. N., Cheong, J. L. Y., Dalach, P., Doyle, L. W., Loke, Y. J., et al. (2016). An epigenetic clock for gestational age at birth based on blood methylation data. *Genome Biology, 17*(1), 206. https://doi.org/10.1186/s13059-016-1068-z

Knudsen, E. I., Knudsen, P. F., & Esterly, S. D. (1984). A critical period for the recovery of sound localization accuracy following monaural occlusion in the barn owl. *The Journal of Neuroscience: The Official Journal of the Society for Neuroscience, 4*(4), 1012–1020.

Kobayashi, C., Glover, G. H., & Temple, E. (2006). Cultural and linguistic influence on neural bases of "Theory of Mind": An fMRI study with Japanese bilinguals. *Brain and Language, 98*(2), 210–220. https://doi.org/10.1016/j.bandl.2006.04.013

Kobayashi, C., Glover, G. H., & Temple, E. (2007). Cultural and linguistic effects on neural bases of "Theory of Mind" in American and Japanese children. *Brain Research, 1164*, 95–107. https://doi.org/10.1016/j.brainres.2007.06.022

Koch, C. (1996). A neuronal correlate of consciousness? *Current Biology: CB, 6*(5), 492. https://doi.org/10.1016/s0960-9822(02)00519-5

Koch, C. (2019). *The feeling of life itself: Why consciousness is widespread but can't be computed.* MIT Press.

Koch, C. M., & Wagner, W. (2011). Epigenetic-aging-signature to determine age in different tissues. *Aging, 3*(10), 1018–1027. https://doi.org/10.18632/aging.100395

Koelsch, S., Schröger, E., & Tervaniemi, M. (1999). Superior pre-attentive auditory processing in musicians. *Neuroreport, 10*(6), 1309–1313. https://doi.org/10.1097/00001756-199904260-00029

Koelsche, C., Schrimpf, D., Stichel, D., Sill, M., Sahm, F., Reuss, D. E., Blattner, M., Worst, B., Heilig, C. E., Beck, K., Horak, P., Kreutzfeldt, S., Paff, E., Stark, S., Johann, P., Selt, F., Ecker, J., Sturm, D., Pajtler, K. W., et al. (2021). Sarcoma classification by DNA methylation profiling. *Nature Communications, 12*(1), Article 1. https://doi.org/10.1038/s41467-020-20603-4

Kohler, E., Keysers, C., Umiltà, M. A., Fogassi, L., Gallese, V., & Rizzolatti, G. (2002). Hearing sounds, understanding actions: Action representation in mirror neurons. *Science (New York, N.Y.), 297*(5582), 846–848. https://doi.org/10.1126/science.1070311

Kolber, A. J. (2016). Free will as a matter of law. In D. Patterson & M. S. Pardo (Eds.), *Philosophical foundations of law and neuroscience* (p. 0). Oxford University Press. https://doi.org/10.1093/acprof:oso/9780198743095.003.0002

Koldobskiy, M. A., Jenkinson, G., Abante, J., DiBlasi, V. A. R., Zhou, W., Pujadas, E., Idrizi, A., Tryggvadottir, R., Callahan, C., Bonifant, C. L., Rabin, K. R., Brown, P. A., Ji, H., Goutsias, J., & Feinberg, A. P. (2021). An information-theory analysis of DNA methylation identifies converging genetic and epigenetic drivers of paediatric acute lymphoblastic leukaemia. *Nature Biomedical Engineering, 5*(4), 360–376. https://doi.org/10.1038/s41551-021-00703-2

Kriegeskorte, N., & Golan, T. (2019). Neural network models and deep learning. *Current Biology, 29*(7), R231–R236. https://doi.org/10.1016/j.cub.2019.02.034

Krishna Temburni, M., & Jacob, M. H. (2001). New functions for glia in the brain. *Proceedings of the National Academy of Sciences, 98*(7), 3631–3632. https://doi.org/10.1073/pnas.081073198

Krüger, M., & Lux, V. (2023). Failure of motor function—A Developmental Embodiment Research perspective on the systemic effects of stress. *Frontiers in Human Neuroscience, 17*, 1083200. https://www.frontiersin.org/articles/10.3389/fnhum.2023.1083200

Kullmann, D. M., & Lamsa, K. P. (2007). Long-term synaptic plasticity in hippocampal interneurons. *Nature Reviews. Neuroscience, 8*(9), 687–699. https://doi.org/10.1038/nrn2207

Kumsta, R. (2019). The role of epigenetics for understanding mental health difficulties and its implications for psychotherapy research. *Psychology and Psychotherapy: Theory, Research and Practice, 92*, 190–207. https://doi.org/10.1111/papt.12227

Kundaje, A., Meuleman, W., Ernst, J., Bilenky, M., Yen, A., Heravi-Moussavi, A., Kheradpour, P., Zhang, Z., Wang, J., Ziller, M. J., Amin, V., Whitaker, J. W., Schultz, M. D., Ward, L. D., Sarkar, A., Quon, G., Sandstrom, R. S., Eaton, M. L., Wu, Y.-C., et al. (2015). Integrative analysis of 111 reference human epigenomes. *Nature, 518*(7539), Article 7539. https://doi.org/10.1038/nature14248

Kyrke-Smith, M., & Williams, J. M. (2018). Bridging synaptic and epigenetic maintenance mechanisms of the engram. *Frontiers in Molecular Neuroscience, 11*, 369. https://www.frontiersin.org/articles/10.3389/fnmol.2018.00369

Labonté, B., Suderman, M., Maussion, G., Lopez, J. P., Navarro-Sánchez, L., Yerko, V., Mechawar, N., Szyf, M., Meaney, M. J., & Turecki, G. (2013). Genome-wide methylation changes in the brains of suicide completers. *The American Journal of Psychiatry, 170*(5), 511–520. https://doi.org/10.1176/appi.ajp.2012.12050627

Laming, D. (2001). Statistical information, uncertainty, and Bayes' theorem: Some applications in experimental psychology. In S. Benferhat & P. Besnard (Eds.), *Symbolic and quantitative approaches to reasoning with uncertainty* (pp. 635–646). Springer. https://doi.org/10.1007/3 -540-44652-4_56

Lancaster, M. A., Renner, M., Martin, C.-A., Wenzel, D., Bicknell, L. S., Hurles, M. E., Homfray, T., Penninger, J. M., Jackson, A. P., & Knoblich, J. A. (2013). Cerebral organoids model human brain development and microcephaly. *Nature, 501*(7467), 373–379. https://doi.org/10.1038/ nature12517

Landecker, H. (2005, January 11). Living differently in time: Plasticity, temporality and cellular biotechnologies. *Culture Machine.* https://culturemachine.net/biopolitics/ living-differently-in-time/

Landecker, H. (2007). *Culturing life: How cells became technologies.* Harvard University Press. https://hdl.handle.net/2027/heb09113.0001.001

Lappalainen, T., & Greally, J. M. (2017). Associating cellular epigenetic models with human phenotypes. *Nature Reviews Genetics, 18*(7), Article 7. https://doi.org/10.1038/nrg.2017.32

Laszlo, A. H., Derrington, I. M., Brinkerhoff, H., Langford, K. W., Nova, I. C., Samson, J. M., Bartlett, J. J., Pavlenok, M., & Gundlach, J. H. (2013). Detection and mapping of 5-methylcytosine and 5-hydroxymethylcytosine with nanopore MspA. *Proceedings of the National Academy of Sciences, 110*(47), 18904–18909. https://doi.org/10.1073/pnas.1310240110

Laurent, L., Wong, E., Li, G., Huynh, T., Tsirigos, A., Ong, C. T., Low, H. M., Kin Sung, K. W., Rigoutsos, I., Loring, J., & Wei, C.-L. (2010). Dynamic changes in the human methylome during differentiation. *Genome Research, 20*(3), 320–331. https://doi.org/10.1101/gr.101907.109

Law, P.-P., & Holland, M. L. (2019). DNA methylation at the crossroads of gene and environment interactions. *Essays in Biochemistry, 63*(6), 717–726. https://doi.org/10.1042/EBC20190031

Lazaris, C., Aifantis, I., & Tsirigos, A. (2020). On epigenetic plasticity and genome topology. *Trends in Cancer, 6*(3), 177–180. https://doi.org/10.1016/j.trecan.2020.01.006

Leontiev, A. N. (1981). *Problems of the development of the mind.* Progress Publ.

Lerner, R. M., & Overton, W. F. (2017). Reduction to absurdity: Why epigenetics invalidates all models involving genetic reduction. *Human Development, 60*(2/3), 107–123.

Levenson, J. M., & Sweatt, J. D. (2005). Epigenetic mechanisms in memory formation. *Nature Reviews Neuroscience, 6*(2), Article 2. https://doi.org/10.1038/nrn1604

Levenson, J. M., O'Riordan, K. J., Brown, K. D., Trinh, M. A., Molfese, D. L., & Sweatt, J. D. (2004). Regulation of histone acetylation during memory formation in the hippocampus *. *Journal of Biological Chemistry, 279*(39), 40545–40559. https://doi.org/10.1074/jbc. M402229200

Levenson, J. M., Roth, T. L., Lubin, F. D., Miller, C. A., Huang, I.-C., Desai, P., Malone, L. M., & Sweatt, J. D. (2006). Evidence that DNA (cytosine-5) methyltransferase regulates synaptic plasticity in the hippocampus. *The Journal of Biological Chemistry, 281*(23), 15763–15773. https://doi.org/10.1074/jbc.M511767200

Levine, M. E., Lu, A. T., Quach, A., Chen, B. H., Assimes, T. L., Bandinelli, S., Hou, L., Baccarelli, A. A., Stewart, J. D., Li, Y., Whitsel, E. A., Wilson, J. G., Reiner, A. P., Aviv, A., Lohman, K., Liu, Y., Ferrucci, L., & Horvath, S. (2018). An epigenetic biomarker of aging for lifespan and healthspan. *Aging (Albany NY), 10*(4), 573–591. https://doi.org/10.18632/aging.101414

Levitan, I. B., & Kaczmarek, L. K. (2015). *The neuron: Cell and molecular biology* (4th ed.). Oxford University Press.

Li, A., Koch, Z., & Ideker, T. (2022a). Epigenetic aging: Biological age prediction and informing a mechanistic theory of aging. *Journal of Internal Medicine, 292*(5), 733–744. https://doi. org/10.1111/joim.13533

Li, A., Mueller, A., English, B., Arena, A., Vera, D., Kane, A. E., & Sinclair, D. A. (2022b). Novel feature selection methods for construction of accurate epigenetic clocks. *PLoS Computational Biology, 18*(8), e1009938. https://doi.org/10.1371/journal.pcbi.1009938

Li, S., Ye, Z., Mather, K. A., Nguyen, T. L., Dite, G. S., Armstrong, N. J., Wong, E. M., Thalamuthu, A., Giles, G. G., Craig, J. M., Saffery, R., Southey, M. C., Tan, Q., Sachdev, P. S.,

& Hopper, J. L. (2022c). Early life affects late-life health through determining DNA methylation across the lifespan: A twin study. *eBioMedicine, 77*, 103927. https://doi.org/10.1016/j.ebiom.2022.103927

Liang, L., Chang, Y., Lu, J., Wu, X., Liu, Q., Zhang, W., Su, X., & Zhang, B. (2019). Global methylomic and transcriptomic analyses reveal the broad participation of DNA methylation in daily gene expression regulation of Populus trichocarpa. *Frontiers in Plant Science, 10*, 243. https://doi.org/10.3389/fpls.2019.00243

Liberti, W. A., Markowitz, J. E., Perkins, L. N., Liberti, D. C., Leman, D. P., Guitchounts, G., Velho, T., Kotton, D. N., Lois, C., & Gardner, T. J. (2016). Unstable neurons underlie a stable learned behavior. *Nature Neuroscience, 19*(12), 1665–1671. https://doi.org/10.1038/nn.4405

Libet, B. W. (1999). Do we have free will? *Journal of Consciousness Studies, 6*(8–9), 47–57.

Libet, B. W., Gleason, C. A., Wright, E. W., & Pearl, D. K. (1983). Time of conscious intention to act in relation to onset of cerebral activity (readiness-potential). The unconscious initiation of a freely voluntary act. *Brain: A Journal of Neurology, 106*(Pt 3), 623–642. https://doi.org/10.1093/brain/106.3.623

Lister, R., Pelizzola, M., Dowen, R. H., Hawkins, R. D., Hon, G., Tonti-Filippini, J., Nery, J. R., Lee, L., Ye, Z., Ngo, Q.-M., Edsall, L., Antosiewicz-Bourget, J., Stewart, R., Ruotti, V., Millar, A. H., Thomson, J. A., Ren, B., & Ecker, J. R. (2009). Human DNA methylomes at base resolution show widespread epigenomic differences. *Nature, 462*(7271), 315–322. https://doi.org/10.1038/nature08514

Liu, A. (2010). Laser capture microdissection in the tissue biorepository. *Journal of Biomolecular Techniques: JBT, 21*(3), 120–125.

Liu, Y., Aryee, M. J., Padyukov, L., Fallin, M. D., Hesselberg, E., Runarsson, A., Reinius, L., Acevedo, N., Taub, M., Ronninger, M., Shchetynsky, K., Scheynius, A., Kere, J., Alfredsson, L., Klareskog, L., Ekström, T. J., & Feinberg, A. P. (2013). Epigenome-wide association data implicate DNA methylation as an intermediary of genetic risk in Rheumatoid Arthritis. *Nature Biotechnology, 31*(2), 142–147. https://doi.org/10.1038/nbt.2487

Lobo, M. K., Karsten, S. L., Gray, M., Geschwind, D. H., & Yang, X. W. (2006). FACS-array profiling of striatal projection neuron subtypes in juvenile and adult mouse brains. *Nature Neuroscience, 9*(3), 443–452. https://doi.org/10.1038/nn1654

Lonze, B. E., & Ginty, D. D. (2002). Function and regulation of CREB family transcription factors in the nervous system. *Neuron, 35*(4), 605–623. https://doi.org/10.1016/S0896-6273(02)00828-0

Lowe, R., & Rakyan, V. K. (2014). Correcting for cell-type composition bias in epigenome-wide association studies. *Genome Medicine, 6*(3), 23. https://doi.org/10.1186/gm540

Lu, A. T., Quach, A., Wilson, J. G., Reiner, A. P., Aviv, A., Raj, K., Hou, L., Baccarelli, A. A., Li, Y., Stewart, J. D., Whitsel, E. A., Assimes, T. L., Ferrucci, L., & Horvath, S. (2019). DNA methylation GrimAge strongly predicts lifespan and healthspan. *Aging, 11*(2), 303–327. https://doi.org/10.18632/aging.101684

Luce, R. D. (2003). Whatever happened to information theory in psychology? *Review of General Psychology, 7*, 183–188. https://doi.org/10.1037/1089-2680.7.2.183

Lux, V. (2013). With Gottlieb beyond Gottlieb: The role of epigenetics in psychobiological development. *International Journal of Developmental Science, 7*(2), 69–78. https://doi.org/10.3233/DEV-1300073

Lux, V. (2016). Conrad Hal Waddingtons "Chreode". In *Synergie: Kultur- und Wissensgeschichte einer Denkfigur; Trajekte* (pp. 247–263). Wilhelm Fink. http://publikationen.ub.uni-frankfurt.de/frontdoor/index/index/docId/46936

Lux, V. (2018). Epigenetic programming effects of early life stress: A dual-activation hypothesis. *Current Genomics, 19*(8), 638–652. https://doi.org/10.2174/1389202919666180307151358

Lux, V., & Richter, J. (Eds.). (2014a). *Kulturen der Epigenetik: Vererbt, codiert, übertragen*. De Gruyter.

Lux, V., & Richter, J. T. (2014b). Einleitung. In *Kulturen der Epigenetik: Vererbt, codiert, übertragen* (pp. xiii–xxviii). De Gruyter. https://doi.org/10.1515/9783110316032.xiii

Lux, V., Non, A. L., Pexman, P. M., Stadler, W., Weber, L. A. E., & Krüger, M. (2021). A developmental framework for embodiment research: The next step toward integrating concepts and methods. *Frontiers in Systems Neuroscience, 15*, 672740. https://doi.org/10.3389/fnsys.2021.672740

Ma, D. K., Marchetto, M. C., Guo, J. U., Ming, G., Gage, F. H., & Song, H. (2010). Epigenetic choreographers of neurogenesis in the adult mammalian brain. *Nature Neuroscience, 13*(11), 1338–1344. https://doi.org/10.1038/nn.2672

MacKay, D. M., & McCulloch, W. S. (1952). The limiting information capacity of a neuronal link. *The Bulletin of Mathematical Biophysics, 14*(2), 127–135. https://doi.org/10.1007/BF02477711

Mackey, A. P., Singley, A. T. M., & Bunge, S. A. (2013). Intensive reasoning training alters patterns of brain connectivity at rest. *Journal of Neuroscience, 33*(11), 4796–4803. https://doi.org/10.1523/JNEUROSCI.4141-12.2013

Maguire, E. A., Gadian, D. G., Johnsrude, I. S., Good, C. D., Ashburner, J., Frackowiak, R. S. J., & Frith, C. D. (2000). Navigation-related structural change in the hippocampi of taxi drivers. *Proceedings of the National Academy of Sciences, 97*(8), 4398–4403. https://doi.org/10.1073/pnas.070039597

Maiers, W. (2009). Conceptual confusions in understanding human action and experience. In T. Teo, P. Stenner, A. Rutherford, E. Park, & C. Baerveldt (Eds.), *Varieties of theoretical psychology: International philosophical and practical concerns* (pp. 101–112). Captus University Publications.

Malik, W. Q., & Ajemian, R. (2017). *Microarrays in the brain: Can they be used for brain-machine interface control?* (pp. 3–39). https://doi.org/10.1016/B978-0-12-800454-8.00001-X

Manuck, S. B. (2010). The reaction norm in gene-environment interaction. *Molecular Psychiatry, 15*(9), 881–882. https://doi.org/10.1038/mp.2009.139

Margolis, D. J., Lütcke, H., Schulz, K., Haiss, F., Weber, B., Kügler, S., Hasan, M. T., & Helmchen, F. (2012). Reorganization of cortical population activity imaged throughout long-term sensory deprivation. *Nature Neuroscience, 15*(11), Article 11. https://doi.org/10.1038/nn.3240

Marshall, P., & Bredy, T. W. (2016). Cognitive neuroepigenetics: The next evolution in our understanding of the molecular mechanisms underlying learning and memory? *Npj Science of Learning, 1*(1), Article 1. https://doi.org/10.1038/npjscilearn.2016.14

Martin, R. (2019). *Neuroscience methods: A guide for advanced students*. CRC Press. https://doi.org/10.1201/9780367810665

Martins, J., Czamara, D., Sauer, S., Rex-Haffner, M., Dittrich, K., Dörr, P., de Punder, K., Overfeld, J., Knop, A., Dammering, F., Entringer, S., Winter, S. M., Buss, C., Heim, C., & Binder, E. B. (2021). Childhood adversity correlates with stable changes in DNA methylation trajectories in children and converges with epigenetic signatures of prenatal stress. *Neurobiology of Stress, 15*, 100336. https://doi.org/10.1016/j.ynstr.2021.100336

Mataró, M., Jurado, M. A., García-Sánchez, C., Barraquer, L., Costa-Jussà, F. R., & Junqué, C. (2001). Long-term effects of bilateral frontal brain lesion: 60 years after injury with an iron bar. *Archives of Neurology, 58*(7), 1139–1142. https://doi.org/10.1001/archneur.58.7.1139

Mau, W., Hasselmo, M. E., & Cai, D. J. (2020). The brain in motion: How ensemble fluidity drives memory-updating and flexibility. *eLife, 9*, e63550. https://doi.org/10.7554/eLife.63550

McGowan, P. O., Sasaki, A., D'Alessio, A. C., Dymov, S., Labonté, B., Szyf, M., Turecki, G., & Meaney, M. J. (2009). Epigenetic regulation of the glucocorticoid receptor in human brain associates with childhood abuse. *Nature Neuroscience, 12*(3), 342–348. https://doi.org/10.1038/nn.2270

McGregor, K., Bernatsky, S., Colmegna, I., Hudson, M., Pastinen, T., Labbe, A., & Greenwood, C. M. T. (2016). An evaluation of methods correcting for cell-type heterogeneity in DNA methylation studies. *Genome Biology, 17*(1), 84. https://doi.org/10.1186/s13059-016-0935-y

McKenna, T. M., McMullen, T. A., & Shlesinger, M. F. (1994). The brain as a dynamic physical system. *Neuroscience, 60*(3), 587–605. https://doi.org/10.1016/0306-4522(94)90489-8

McKenzie, A. T., Wang, M., Hauberg, M. E., Fullard, J. F., Kozlenkov, A., Keenan, A., Hurd, Y. L., Dracheva, S., Casaccia, P., Roussos, P., & Zhang, B. (2018). Brain cell type specific gene expression and co-expression network architectures. *Scientific Reports, 8*(1), Article 1. https://doi.org/10.1038/s41598-018-27293-5

McMullen, S., & Mostyn, A. (2009). Animal models for the study of the developmental origins of health and disease: Workshop on "Nutritional models of the developmental origins of adult health and disease". *Proceedings of the Nutrition Society, 68*(3), 306–320. https://doi.org/10.1017/S0029665109001396

Mermillod, M., Bugaiska, A., & Bonin, P. (2013). The stability-plasticity dilemma: Investigating the continuum from catastrophic forgetting to age-limited learning effects. *Frontiers in Psychology, 4*, 504. https://www.frontiersin.org/articles/10.3389/fpsyg.2013.00504

Merzenich, M. M. (2013). *Soft-wired: How the new science of brain plasticity can change your life* (2nd ed.). Parnassus Publ.

Merzenich, M. M., & DeCharms, R. C. (1996). Neural representations, experience, and change. In R. Llinas & P. Chruchland (Eds.), *The mind-brain continuum* (pp. 61–81). MIT Press.

Merzenich, M. M., & Jenkins, W. M. (1993). Reorganization of cortical representations of the hand following alterations of skin inputs induced by nerve injury, skin island transfers, and experience. *Journal of Hand Therapy: Official Journal of the American Society of Hand Therapists, 6*(2), 89–104. https://doi.org/10.1016/s0894-1130(12)80290-0

Mews, P., Calipari, E. S., Day, J., Lobo, M. K., Bredy, T., & Abel, T. (2021). From circuits to chromatin: The emerging role of epigenetics in mental health. *Journal of Neuroscience, 41*(5), 873–882. https://doi.org/10.1523/JNEUROSCI.1649-20.2020

Miller, E. K., & Cohen, J. D. (2001). An integrative theory of prefrontal cortex function. *Annual Review of Neuroscience, 24*, 167–202. https://doi.org/10.1146/annurev.neuro.24.1.167

Miller, C. A., & Sweatt, J. D. (2007). Covalent modification of DNA regulates memory formation. *Neuron, 53*(6), 857–869. https://doi.org/10.1016/j.neuron.2007.02.022

Miller, C. A., Campbell, S. L., & Sweatt, J. D. (2008). DNA methylation and histone acetylation work in concert to regulate memory formation and synaptic plasticity. *Neurobiology of Learning and Memory, 89*(4), 599–603. https://doi.org/10.1016/j.nlm.2007.07.016

Millman, D. J., Ocker, G. K., Caldejon, S., Kato, I., Larkin, J. D., Lee, E. K., Luviano, J., Nayan, C., Nguyen, T. V., North, K., Seid, S., White, C., Lecoq, J., Reid, C., Buice, M. A., & de Vries, S. E. (2020). VIP interneurons in mouse primary visual cortex selectively enhance responses to weak but specific stimuli. *eLife, 9*, e55130. https://doi.org/10.7554/eLife.55130

Minatohara, K., Akiyoshi, M., & Okuno, H. (2016). Role of immediate-early genes in synaptic plasticity and neuronal ensembles underlying the memory trace. *Frontiers in Molecular Neuroscience, 8*, 78. https://www.frontiersin.org/articles/10.3389/fnmol.2015.00078

Mischel, T. (1970). Wundt and the conceptual foundations of psychology. *Philosophy and Phenomenological Research, 31*(1), 1–26. https://doi.org/10.2307/2105977

Morales Berstein, F., McCartney, D. L., Lu, A. T., Tsilidis, K. K., Bouras, E., Haycock, P., Burrows, K., Phipps, A. I., Buchanan, D. D., Cheng, I., Martin, R. M., Davey Smith, G., Relton, C. L., Horvath, S., Marioni, R. E., Richardson, T. G., Richmond, R. C., & the PRACTICAL consortium. (2022). Assessing the causal role of epigenetic clocks in the development of multiple cancers: A Mendelian randomization study. *eLife, 11*, e75374. https://doi.org/10.7554/eLife.75374

Moroz, L. L. (2018). NeuroSystematics and periodic system of neurons: Model vs reference species at single-cell resolution. *ACS Chemical Neuroscience, 9*(8), 1884–1903. https://doi.org/10.1021/acschemneuro.8b00100

Moroz, L. L. (2021). Multiple origins of neurons from secretory cells. *Frontiers in Cell and Developmental Biology, 9*, 669087. https://www.frontiersin.org/articles/10.3389/fcell.2021.669087

Mott, D. D., & Dingledine, R. (2003). Interneuron diversity series: Interneuron research – Challenges and strategies. *Trends in Neurosciences, 26*(9), 484–488. https://doi.org/10.1016/S0166-2236(03)00200-5

Mountcastle, V. B. (1957). Modality and topographic properties of single neurons of cat's somatic sensory cortex. *Journal of Neurophysiology, 20*(4), 408–434. https://doi.org/10.1152/jn.1957.20.4.408

Mozzachiodi, R., & Byrne, J. H. (2010). More than synaptic plasticity: Role of nonsynaptic plasticity in learning and memory. *Trends in Neurosciences, 33*(1), 17. https://doi.org/10.1016/j.tins.2009.10.001

Muñoz-Martin, I., Bianchi, S., Hashemkhani, S., Pedretti, G., Melnic, O., & Ielmini, D. (2021). A brain-inspired homeostatic neuron based on phase-change memories for efficient neuromorphic computing. *Frontiers in Neuroscience, 15*, 709053. https://www.frontiersin.org/articles/10.3389/fnins.2021.709053

Münte, T. F., Altenmüller, E., & Jäncke, L. (2002). The musician's brain as a model of neuroplasticity. *Nature Reviews Neuroscience, 3*(6), Article 6. https://doi.org/10.1038/nrn843

Murgatroyd, C., Patchev, A. V., Wu, Y., Micale, V., Bockmühl, Y., Fischer, D., Holsboer, F., Wotjak, C. T., Almeida, O. F. X., & Spengler, D. (2009). Dynamic DNA methylation programs persistent adverse effects of early-life stress. *Nature Neuroscience, 12*(12), 1559–1566. https://doi.org/10.1038/nn.2436

Murphy, N. C., Ellis, G. F. R., & O'Connor, T. (Eds.). (2009). *Downward causation and the neurobiology of free will*. Springer Verlag.

Nagelhus, E. A., Amiry-Moghaddam, M., Bergersen, L. H., Bjaalie, J. G., Eriksson, J., Gundersen, V., Leergaard, T. B., Morth, J. P., Storm-Mathisen, J., Torp, R., Walhovd, K. B., & Tønjum, T. (2013). The glia doctrine: Addressing the role of glial cells in healthy brain ageing. *Mechanisms of Ageing and Development, 134*(10), 449–459. https://doi.org/10.1016/j.mad.2013.10.001

Nagy, C., & Turecki, G. (2012). Sensitive periods in epigenetics: Bringing us closer to complex behavioral phenotypes. *Epigenomics, 4*(4), 445–457. https://doi.org/10.2217/epi.12.37

Nahum, M., Lee, H., & Merzenich, M. M. (2013). Principles of neuroplasticity-based rehabilitation. *Progress in Brain Research, 207*, 141–171. https://doi.org/10.1016/B978-0-444-63327-9.00009-6

Naumova, O. Y., Lee, M., Rychkov, S. Y., Vlasova, N. V., & Grigorenko, E. L. (2013). Gene expression in the human brain: The current state of the study of specificity and spatio-temporal dynamics. *Child Development, 84*(1), 76–88. https://doi.org/10.1111/cdev.12014

Negi, S. K., & Guda, C. (2017). Global gene expression profiling of healthy human brain and its application in studying neurological disorders. *Scientific Reports, 7*(1), Article 1. https://doi.org/10.1038/s41598-017-00952-9

Nicolelis, M. A. L., Ghazanfar, A. A., Faggin, B. M., Votaw, S., & Oliveira, L. M. O. (1997). Reconstructing the engram: Simultaneous, multisite, many single neuron recordings. *Neuron, 18*(4), 529–537. https://doi.org/10.1016/S0896-6273(00)80295-0

Niu, Y., DesMarais, T. L., Tong, Z., Yao, Y., & Costa, M. (2015). Oxidative stress alters global histone modification and DNA methylation. *Free Radical Biology and Medicine, 82*, 22–28. https://doi.org/10.1016/j.freeradbiomed.2015.01.028

Norris, G. T., Smirnov, I., Filiano, A. J., Shadowen, H. M., Cody, K. R., Thompson, J. A., Harris, T. H., Gaultier, A., Overall, C. C., & Kipnis, J. (2018). Neuronal integrity and complement control synaptic material clearance by microglia after CNS injury. *The Journal of Experimental Medicine, 215*(7), 1789–1801. https://doi.org/10.1084/jem.20172244

Northoff, G. (2023). *Neuropsychoanalysis: A contemporary introduction*. Routledge.

Nshimiyimana, J., Uwihoreye, P., Muhigirwa, J. C., Niyonsega, T., Nshimiyimana, J., Uwihoreye, P., Muhigirwa, J. C., & Niyonsega, T. (2023). Neurofunctional intervention approaches. In *Neurorehabilitation and physical therapy*. IntechOpen. https://doi.org/10.5772/intechopen.106604

O'Neill, H., Lee, H., Gupta, I., Rodger, E. J., & Chatterjee, A. (2022). Single-cell DNA methylation analysis in cancer. *Cancers, 14*(24), Article 24. https://doi.org/10.3390/cancers14246171

Oblak, L., van der Zaag, J., Higgins-Chen, A. T., Levine, M. E., & Boks, M. P. (2021). A systematic review of biological, social and environmental factors associated with epigenetic clock acceleration. *Ageing Research Reviews, 69*, 101348. https://doi.org/10.1016/j.arr.2021.101348

Odling-Smee, F. J., Laland, K. N., & Feldman, M. W. (2003). *Niche construction: The neglected process in evolution* (Online-ausg). Princeton University Press.

Oh, E. S., & Petronis, A. (2021). Origins of human disease: The chrono-epigenetic perspective. *Nature Reviews Genetics, 22*(8), Article 8. https://doi.org/10.1038/s41576-021-00348-6

Oh, G., Ebrahimi, S., Carlucci, M., Zhang, A., Nair, A., Groot, D. E., Labrie, V., Jia, P., Oh, E. S., Jeremian, R. H., Susic, M., Shrestha, T. C., Ralph, M. R., Gordevičius, J., Koncevičius, K., & Petronis, A. (2018). Cytosine modifications exhibit circadian oscillations that are involved in epigenetic diversity and aging. *Nature Communications, 9*(1), Article 1. https://doi.org/10.1038/s41467-018-03073-7

Oh, G., Koncevičius, K., Ebrahimi, S., Carlucci, M., Groot, D. E., Nair, A., Zhang, A., Kriščiūnas, A., Oh, E. S., Labrie, V., Wong, A. H. C., Gordevičius, J., Jia, P., Susic, M., & Petronis, A. (2019). Circadian oscillations of cytosine modification in humans contribute to epigenetic variability, aging, and complex disease. *Genome Biology, 20*(1), 2. https://doi.org/10.1186/s13059-018-1608-9

Okano, M., Xie, S., & Li, E. (1998). Cloning and characterization of a family of novel mammalian DNA (cytosine-5) methyltransferases. *Nature Genetics, 19*(3), 219–220. https://doi.org/10.1038/890

Olde Loohuis, N. F. M., Kos, A., Martens, G. J. M., Van Bokhoven, H., Nadif Kasri, N., & Aschrafi, A. (2012). MicroRNA networks direct neuronal development and plasticity. *Cellular and Molecular Life Sciences, 69*(1), 89–102. https://doi.org/10.1007/s00018-011-0788-1

Önder, Ö., Sidoli, S., Carroll, M., & Garcia, B. A. (2015). Progress in epigenetic histone modification analysis by mass spectrometry for clinical investigations. *Expert Review of Proteomics, 12*(5), 499–517. https://doi.org/10.1586/14789450.2015.1084231

Oyama, S. (2000). *The ontogeny of information: Developmental systems and evolution* (2nd ed., rev. and expanded). Duke University Press.

Palma-Gudiel, H., Córdova-Palomera, A., Leza, J. C., & Fañanás, L. (2015). Glucocorticoid receptor gene (NR3C1) methylation processes as mediators of early adversity in stress-related disorders causality: A critical review. *Neuroscience and Biobehavioral Reviews, 55*, 520–535. https://doi.org/10.1016/j.neubiorev.2015.05.016

Paninski, L., Ahmadian, Y., Ferreira, D. G., Koyama, S., Rad, K. R., Vidne, M., Vogelstein, J., & Wu, W. (2010). A new look at state-space models for neural data. *Journal of Computational Neuroscience, 29*(0), 107–126. https://doi.org/10.1007/s10827-009-0179-x

Parisi, G. I., Kemker, R., Part, J. L., Kanan, C., & Wermter, S. (2019). Continual lifelong learning with neural networks: A review. *Neural Networks, 113*, 54–71. https://doi.org/10.1016/j.neunet.2019.01.012

Parnes, O. (2007). Die Topographie der Vererbung. Epigenetische Landschaften bei Waddington und Piper. *Trajekte, 14*, 26–31.

Pascual, M., Ibáñez, F., & Guerri, C. (2019). Exosomes as mediators of neuron-glia communication in neuroinflammation. *Neural Regeneration Research, 15*(5), 796–801. https://doi.org/10.4103/1673-5374.268893

Pauen, M. (2007). Self-determination free will, responsibility, and determinism. *Synthesis Philosophica, 22*, 455–475+510.

Paul, D. S., Teschendorff, A. E., Dang, M. A. N., Lowe, R., Hawa, M. I., Ecker, S., Beyan, H., Cunningham, S., Fouts, A. R., Ramelius, A., Burden, F., Farrow, S., Rowlston, S., Rehnstrom, K., Frontini, M., Downes, K., Busche, S., Cheung, W. A., Ge, B., et al. (2016). Increased DNA methylation variability in type 1 diabetes across three immune effector cell types. *Nature Communications, 7*(1), Article 1. https://doi.org/10.1038/ncomms13555

Pawela, C., & Biswal, B. (2011). Brain connectivity: A new journal emerges. *Brain Connectivity, 1*(1), 1–2. https://doi.org/10.1089/brain.2011.0020

Pellicano, A., Mingoia, G., Ritter, C., Buccino, G., & Binkofski, F. (2021). Respiratory function modulated during execution, observation, and imagination of walking via SII. *Scientific Reports, 11*(1), Article 1. https://doi.org/10.1038/s41598-021-03147-5

Penfield, W., & Boldrey, E. (1937). Somatic motor and sensory representation in the cerebral cortex of man as studied by electrical stimulation. *Brain: A Journal of Neurology, 60*, 389–443. https://doi.org/10.1093/brain/60.4.389

Penfield, W., & Rasmussen, T. (1950). *The cerebral cortex of man; a clinical study of localization of function* (pp. xv, 248). Macmillan.

Peng, H., Xie, P., Liu, L., Kuang, X., Wang, Y., Qu, L., Gong, H., Jiang, S., Li, A., Ruan, Z., Ding, L., Yao, Z., Chen, C., Chen, M., Daigle, T. L., Dalley, R., Ding, Z., Duan, Y., Feiner, A., et al. (2021). Morphological diversity of single neurons in molecularly defined cell types. *Nature, 598*(7879), 174–181. https://doi.org/10.1038/s41586-021-03941-1

Perez-Catalan, N. A., Doe, C. Q., & Ackerman, S. D. (2021). The role of astrocyte-mediated plasticity in neural circuit development and function. *Neural Development, 16*(1), 1. https://doi.org/10.1186/s13064-020-00151-9

Peters, A. J., Lee, J., Hedrick, N. G., O'Neil, K., & Komiyama, T. (2017). Reorganization of corticospinal output during motor learning. *Nature Neuroscience, 20*(8), Article 8. https://doi.org/10.1038/nn.4596

Petersen, C. C. H. (2007). The functional organization of the barrel cortex. *Neuron, 56*(2), 339–355. https://doi.org/10.1016/j.neuron.2007.09.017

Pfeifer, R., Bongard, J., & Grand, S. (2007a). *How the body shapes the way we think: A new view of intelligence*. MIT Press.

Pfeifer, R., Lungarella, M., & Iida, F. (2007b). Self-organization, embodiment, and biologically inspired robotics. *Science (New York, N.Y.), 318*(5853), 1088–1093. https://doi.org/10.1126/science.1145803

Piaget, J. (1970). *Science of education and the psychology of the child* (D. Coltman, Trans., p. 186). Orion.

Piaget, J. (1974). *Abriß der genetischen Epistemologie (L'epistémologie génétique, dt. – Übers.:Friitz Kubli. Mit e. Einf. V. Fritz Kubli u.e. Bibliogr. D. Werke v. Piaget.)*.

Piaget, J. (1975). *Die Entwicklung des Erkennens* ([Versch. Aufl.]). Klett.

Piersma, T., & Drent, J. (2003). Phenotypic flexibility and the evolution of organismal design. *Trends in Ecology & Evolution, 18*(5), 228–233. https://doi.org/10.1016/S0169-5347(03)00036-3

Pigliucci, M., Müller, G., & Konrad Lorenz Institute for Evolution and Cognition Research (Eds.). (2010). *Evolution, the extended synthesis*. MIT Press.

Pillow, J. W., Shlens, J., Paninski, L., Sher, A., Litke, A. M., Chichilnisky, E. J., & Simoncelli, E. P. (2008). Spatio-temporal correlations and visual signalling in a complete neuronal population. *Nature, 454*(7207), Article 7207. https://doi.org/10.1038/nature07140

Planques, A., Oliveira Moreira, V., Dubreuil, C., Prochiantz, A., & Di Nardo, A. A. (2019). OTX2 signals from the choroid plexus to regulate adult neurogenesis. *ENeuro, 6*(2), ENEURO.0262-18.2019. https://doi.org/10.1523/ENEURO.0262-18.2019

Pockett, S., Banks, W. P., & Gallagher, S. (2006). *Does consciousness cause behavior?* MIT Press Ebsco Publishing [distributor]. https://search.ebscohost.com/login.aspx?direct=true&scope=site&db=nlebk&db=nlabk&AN=156014

Portin, P., & Wilkins, A. (2017). The evolving definition of the term "Gene". *Genetics, 205*(4), 1353–1364. https://doi.org/10.1534/genetics.116.196956

Price, A. J., Collado-Torres, L., Ivanov, N. A., Xia, W., Burke, E. E., Shin, J. H., Tao, R., Ma, L., Jia, Y., Hyde, T. M., Kleinman, J. E., Weinberger, D. R., & Jaffe, A. E. (2019). Divergent neuronal DNA methylation patterns across human cortical development reveal critical periods and a unique role of CpH methylation. *Genome Biology, 20*(1), 196. https://doi.org/10.1186/s13059-019-1805-1

Prinz, W. (2008). Der Wille als Artefakt. In K.-S. Rehberg (Ed.), *Die Natur der Gesellschaft: Verhandlungen des 33. Kongresses der Deutschen Gesellschaft für Soziologie in Kassel 2006. Teilbd. 1 u. 2* (pp. 642–655). Campus Verl.

Prochiantz, A. (2012). *Qu'est-ce que le vivant?* Éditions du Seuil.

Prochiantz, A., & Di Nardo, A. A. (2015). Homeoprotein signaling in the developing and adult nervous system. *Neuron, 85*(5), 911–925. https://doi.org/10.1016/j.neuron.2015.01.019

Provençal, N., Arloth, J., Cattaneo, A., Anacker, C., Cattane, N., Wiechmann, T., Röh, S., Ködel, M., Klengel, T., Czamara, D., Müller, N. S., Lahti, J., Null, N., Räikkönen, K., Pariante, C. M., Binder, E. B., Kajantie, E., Hämäläinen, E., Villa, P., & Laivuori, H. (2020). Glucocorticoid exposure during hippocampal neurogenesis primes future stress response by inducing changes in DNA methylation. *Proceedings of the National Academy of Sciences, 117*(38), 23280–23285. https://doi.org/10.1073/pnas.1820842116

Prull, M. W., Gabrieli, J. D. E., & Bunge, S. A. (2000). Age-related changes in memory: A cognitive neuroscience perspective. In *The handbook of aging and cognition* (2nd ed., pp. 91–153). Lawrence Erlbaum Associates Publishers.

Ptak, C., & Petronis, A. (2010). Epigenetic approaches to psychiatric disorders. *Dialogues in Clinical Neuroscience, 12*(1), 25–35.

Qi, L., & Teschendorff, A. E. (2022). Cell-type heterogeneity: Why we should adjust for it in epigenome and biomarker studies. *Clinical Epigenetics, 14*(1), 31. https://doi.org/10.1186/s13148-022-01253-3

Qian, X., Nguyen, H. N., Song, M. M., Hadiono, C., Ogden, S. C., Hammack, C., Yao, B., Hamersky, G. R., Jacob, F., Zhong, C., Yoon, K.-J., Jeang, W., Lin, L., Li, Y., Thakor, J., Berg, D. A., Zhang, C., Kang, E., Chickering, M., et al. (2016). Brain-region-specific organoids using mini-bioreactors for modeling ZIKV exposure. *Cell, 165*(5), 1238–1254. https://doi.org/10.1016/j.cell.2016.04.032

Rahman, M. F., & McGowan, P. O. (2022). Cell-type-specific epigenetic effects of early life stress on the brain. *Translational Psychiatry, 12*, 326. https://doi.org/10.1038/s41398-022-02076-9

Rahmani, E., Schweiger, R., Rhead, B., Criswell, L. A., Barcellos, L. F., Eskin, E., Rosset, S., Sankararaman, S., & Halperin, E. (2019). Cell-type-specific resolution epigenetics without the need for cell sorting or single-cell biology. *Nature Communications, 10*(1), Article 1. https://doi.org/10.1038/s41467-019-11052-9

Raichle, M. E. (2009). A paradigm shift in functional brain imaging. *The Journal of Neuroscience, 29*(41), 12729–12734. https://doi.org/10.1523/JNEUROSCI.4366-09.2009

Rajendran, L., Bali, J., Barr, M. M., Court, F. A., Krämer-Albers, E.-M., Picou, F., Raposo, G., van der Vos, K. E., van Niel, G., Wang, J., & Breakefield, X. O. (2014). Emerging roles of extracellular vesicles in the nervous system. *Journal of Neuroscience, 34*(46), 15482–15489. https://doi.org/10.1523/JNEUROSCI.3258-14.2014

Rakic, P. (1981). Neuronal-glial interaction during brain development. *Trends in Neurosciences, 4*, 184–187. https://doi.org/10.1016/0166-2236(81)90060-6

Ramón y Cajal, S. (1893). *Manual de histología normal y técnica micrográfica*. Liberia de Pascual Aguilar. https://wellcomecollection.org/works/wegkqafc

Ramón y Cajal, S. (1900). *Studien über die Hirnrinde des Menschen*. Verlag von Johann Ambrosius Barth.

Ramón y Cajal, S. (1906). *The structure and connexions of neurons*. https://www.nobelprize.org/uploads/2018/06/cajal-lecture.pdf

Rand, A. C., Jain, M., Eizenga, J. M., Musselman-Brown, A., Olsen, H. E., Akeson, M., & Paten, B. (2017). Mapping DNA methylation with high-throughput nanopore sequencing. *Nature Methods, 14*(4), Article 4. https://doi.org/10.1038/nmeth.4189

Rees, T. (2016). *Plastic reason: An anthropology of brain science in embryogenetic terms*. University of California Press.

Reinius, L. E., Acevedo, N., Joerink, M., Pershagen, G., Dahlén, S.-E., Greco, D., Söderhäll, C., Scheynius, A., & Kere, J. (2012). Differential DNA methylation in purified human blood cells: Implications for cell lineage and studies on disease susceptibility. *PLoS One, 7*(7), e41361. https://doi.org/10.1371/journal.pone.0041361

Retzius, G. (1890). *Zur Kenntniss des Nervensystems der Crustaceen* (Vol. I). Samson & Wallin.

Rheinberger, H.-J. (1997). *Toward a history of epistemic things: Synthesizing proteins in the test tube*. Stanford University Press.

Rheinberger, H.-J. (2000). Beyond nature and culture: Modes of reasoning in the age of molecular biology and medicine. In A. Cambrosio, A. Young, & M. Lock (Eds.), *Living and working with the new medical technologies: Intersections of inquiry* (pp. 19–30). Cambridge University Press. https://doi.org/10.1017/CBO9780511621765.002

Rheinberger, H.-J. (2010). *An epistemology of the concrete: Twentieth-century histories of life*. Duke University Press. https://doi.org/10.2307/j.ctv11qdxmc

Ribic, A. (2020). Stability in the face of change: Lifelong experience-dependent plasticity in the sensory cortex. *Frontiers in Cellular Neuroscience, 14*, 76. https://www.frontiersin.org/articles/10.3389/fncel.2020.00076

Riffo-Campos, Á. L., Castillo, J., Tur, G., González-Figueroa, P., Georgieva, E. I., Rodríguez, J. L., López-Rodas, G., Rodrigo, M. I., & Franco, L. (2015). Nucleosome-specific, time-dependent changes in histone modifications during activation of the early growth response 1 (Egr1) gene. *The Journal of Biological Chemistry, 290*(1), 197–208. https://doi.org/10.1074/jbc.M114.579292

Riggs, A. D., Martienssen, R. A., & Russo, V. E. A. (1996). Introduction. In *Epigenetic mechanisms of gene regulation* (Vol. 32, pp. 1–4). Cold Spring Harbor Laboratory Press. https://cshmonographs.org/index.php/monographs/article/view/4519

Ripoli, C. (2017). Engrampigenetics: Epigenetics of engram memory cells. *Behavioural Brain Research, 325*(Pt B), 297–302. https://doi.org/10.1016/j.bbr.2016.11.043

Rivera, C. M., & Ren, B. (2013). Mapping human epigenomes. *Cell, 155*(1), 39–55. https://doi.org/10.1016/j.cell.2013.09.011

Rizzolatti, G., & Craighero, L. (2004). The mirror-neuron system. *Annual Review of Neuroscience, 27*, 169–192. https://doi.org/10.1146/annurev.neuro.27.070203.144230

Rizzolatti, G., & Fogassi, L. (2014). The mirror mechanism: Recent findings and perspectives. *Philosophical Transactions of the Royal Society of London. Series B, Biological Sciences, 369*(1644), 20130420. https://doi.org/10.1098/rstb.2013.0420

Rizzolatti, G., & Sinigaglia, C. (2007). Mirror neurons and motor intentionality. *Functional Neurology, 22*(4), 205–210.

Rizzolatti, G., & Sinigaglia, C. (2016). The mirror mechanism: A basic principle of brain function. *Nature Reviews. Neuroscience, 17*(12), 757–765. https://doi.org/10.1038/nrn.2016.135

Rock, R. B., Gekker, G., Hu, S., Sheng, W. S., Cheeran, M., Lokensgard, J. R., & Peterson, P. K. (2004). Role of microglia in central nervous system infections. *Clinical Microbiology Reviews, 17*(4), 942–964. https://doi.org/10.1128/CMR.17.4.942-964.2004

Roth, G. (2003). *Fühlen, Denken, Handeln: Wie das Gehirn unser Verhalten steuert* (1. Aufl., [Nachdr.]). Suhrkamp.

Ruberti, F., Barbato, C., & Cogoni, C. (2012). Targeting microRNAs in neurons: Tools and perspectives. *Experimental Neurology, 235*(2), 419–426. https://doi.org/10.1016/j.expneurol.2011.10.031

Rulands, S., Lee, H. J., Clark, S. J., Angermueller, C., Smallwood, S. A., Krueger, F., Mohammed, H., Dean, W., Nichols, J., Rugg-Gunn, P., Kelsey, G., Stegle, O., Simons, B. D., & Reik, W. (2018). Genome-scale oscillations in DNA methylation during exit from pluripotency. *Cell Systems, 7*(1), 63–76.e12. https://doi.org/10.1016/j.cels.2018.06.012

Rusconi, F., Grillo, B., Ponzoni, L., Bassani, S., Toffolo, E., Paganini, L., Mallei, A., Braida, D., Passafaro, M., Popoli, M., Sala, M., & Battaglioli, E. (2016). LSD1 modulates stress-evoked transcription of immediate early genes and emotional behavior. *Proceedings of the National Academy of Sciences of the United States of America, 113*(13), 3651–3656. https://doi.org/10.1073/pnas.1511974113

Rüsseler, J., Altenmüller, E., Nager, W., Kohlmetz, C., & Münte, T. F. (2001). Event-related brain potentials to sound omissions differ in musicians and non-musicians. *Neuroscience Letters, 308*(1), 33–36. https://doi.org/10.1016/s0304-3940(01)01977-2

Santhanam, G., Ryu, S. I., Yu, B. M., Afshar, A., & Shenoy, K. V. (2006). A high-performance brain–computer interface. *Nature, 442*(7099), Article 7099. https://doi.org/10.1038/nature04968

Sayood, K. (2018). Information theory and cognition: A review. *Entropy, 20*(9), 706. https://doi.org/10.3390/e20090706

Scheithauer, H., Niebank, K., & Ittel, A. (2009). Developmental science: Integrating knowledge about dynamic processes in human development. In J. Valsiner, P. C. M. Molenaar, M. C. D. P. Lyra, & N. Chaudhary (Eds.), *Dynamic process methodology in the social and developmental sciences* (pp. 595–617). Springer US. https://doi.org/10.1007/978-0-387-95922-1_26

Scheler, G., & Fellous, J.-M. (2001). Dopamine modulation of prefrontal delay activity-reverberatory activity and sharpness of tuning curves. *Neurocomputing, 38–40*, 1549–1556. https://doi.org/10.1016/S0925-2312(01)00559-8

Scherrer, K. (2018). Primary transcripts: From the discovery of RNA processing to current concepts of gene expression – Review. *Experimental Cell Research, 373*(1–2), 1–33. https://doi.org/10.1016/j.yexcr.2018.09.011

Scherrer, K., & Jost, J. (2007). The gene and the genon concept: A functional and information-theoretic analysis. *Molecular Systems Biology, 3*, 87. https://doi.org/10.1038/msb4100123

Schmidt, U., Holsboer, F., & Rein, T. (2011). Epigenetic aspects of posttraumatic stress disorder. *Disease Markers, 30*(2–3), 77–87. https://doi.org/10.3233/DMA-2011-0749

Schmidt-Wilcke, T., Fuchs, E., Funke, K., Vlachos, A., Muller-Dahlhaus, F., Puts, N. A. J., Harris, R. E., & Edden, R. A. E. (2018). GABA-from inhibition to cognition. *The Neuroscientist: A Review Journal Bringing Neurobiology, Neurology and Psychiatry, 24*(5), 501–515. https://doi.org/10.1177/1073858417734530

Schneegans, S., Lins, J., & Schöner, G. (2015). Embedding dynamic field theory in neurophysiology. In G. Schöner, J. Spencer, & D. Research Group (Eds.), *Dynamic thinking: A primer on dynamic field theory* (p. 0). Oxford University Press. https://doi.org/10.1093/acprof:oso/9780199300563.003.0003

Schöner, G. (2014). Dynamical systems thinking: From metaphor to neural theory. In *Handbook of developmental systems theory and methodology* (pp. 188–217). The Guilford Press.

Schöner, G. (2020). The dynamics of neural populations capture the laws of the mind. *Topics in Cognitive Science, 12*(4), 1257–1271. https://doi.org/10.1111/tops.12453

Schöner, G., & Schutte, A. R. (2015). Dynamic field theory: Foundations. In G. Schöner, J. Spencer, & D. Research Group (Eds.), *Dynamic thinking: A primer on dynamic field theory* (p. 0). Oxford University Press. https://doi.org/10.1093/acprof:oso/9780199300563.003.0002

Schöner, G., & Spencer, J. (2015). *Dynamic thinking: A primer on dynamic field theory*. Oxford University Press.

Schöner, G., Reimann, H., & Lins, J. (2015). Neural dynamics. In G. Schöner, J. Spencer, & D. Research Group (Eds.), *Dynamic thinking: A primer on dynamic field theory* (p. 0). Oxford University Press. https://doi.org/10.1093/acprof:oso/9780199300563.003.0001

Schug, S., Benzing, F., & Steger, A. (2021). Presynaptic stochasticity improves energy efficiency and helps alleviate the stability-plasticity dilemma. *eLife, 10*, e69884. https://doi.org/10.7554/eLife.69884

Scott, C. A., Duryea, J. D., MacKay, H., Baker, M. S., Laritsky, E., Gunasekara, C. J., Coarfa, C., & Waterland, R. A. (2020). Identification of cell type-specific methylation signals in bulk whole genome bisulfite sequencing data. *Genome Biology, 21*(1), 156. https://doi.org/10.1186/s13059-020-02065-5

Seidel-Marzi, O., & Ragert, P. (2020). Neurodiagnostics in sports: Investigating the athlete's brain to augment performance and sport-specific skills. *Frontiers in Human Neuroscience, 14*, 133. https://doi.org/10.3389/fnhum.2020.00133

Seligman, R. (2017). "Bio-looping" and the psychophysiological in religious belief and practice: Mechanisms of embodiment in Candomblé trance and possession. In *The Palgrave handbook of biology and society* (pp. 417–439). Palgrave Macmillan. https://doi.org/10.1057/978-1-137-52879-7_18

Seligman, R., Choudhury, S., & Kirmayer, L. J. (2016). Locating culture in the brain and in the world: From social categories to the ecology of mind. In *The Oxford handbook of cultural neuroscience* (pp. 3–20). Oxford University Press.

Sepulcre, J., Liu, H., Talukdar, T., Martincorena, I., Yeo, B. T. T., & Buckner, R. (2010). The organization of local and distant functional connectivity in the human brain. *PLoS Computational Biology, 6*, e1000808. https://doi.org/10.1371/journal.pcbi.1000808

Shadlen, M. N., & Movshon, J. A. (1999). Synchrony unbound: A critical evaluation of the temporal binding hypothesis. *Neuron, 24*(1), 67–77, 111–125. https://doi.org/10.1016/s0896-6273(00)80822-3

Shannon, C. E. (1948). A mathematical theory of communication. *The Bell System Technical Journal, 27*(3), 379–423. https://doi.org/10.1002/j.1538-7305.1948.tb01338.x

Sharpee, T. O. (2014). Toward functional classification of neuronal types. *Neuron, 83*(6), 1329–1334. https://doi.org/10.1016/j.neuron.2014.08.040

Sherrington, C. S. (1906). *The integrative action of the nervous system* (pp. xvi, 411). Yale University Press. https://doi.org/10.1037/13798-000

Shireby, G. L., Davies, J. P., Francis, P. T., Burrage, J., Walker, E. M., Neilson, G. W. A., Dahir, A., Thomas, A. J., Love, S., Smith, R. G., Lunnon, K., Kumari, M., Schalkwyk, L. C., Morgan, K., Brookes, K., Hannon, E., & Mill, J. (2020). Recalibrating the epigenetic clock: Implications for assessing biological age in the human cortex. *Brain, 143*(12), 3763–3775. https://doi.org/10.1093/brain/awaa334

Sidiropoulou, K., Pissadaki, E. K., & Poirazi, P. (2006). Inside the brain of a neuron. *EMBO Reports, 7*(9), 886–892. https://doi.org/10.1038/sj.embor.7400789

Simpkin, A. J., Hemani, G., Suderman, M., Gaunt, T. R., Lyttleton, O., Mcardle, W. L., Ring, S. M., Sharp, G. C., Tilling, K., Horvath, S., Kunze, S., Peters, A., Waldenberger, M., Ward-Caviness, C., Nohr, E. A., Sørensen, T. I. A., Relton, C. L., & Smith, G. D. (2016). Prenatal and early life influences on epigenetic age in children: A study of mother-offspring pairs from two cohort studies. *Human Molecular Genetics, 25*(1), 191–201. https://doi.org/10.1093/hmg/ddv456

Simpkin, A. J., Howe, L. D., Tilling, K., Gaunt, T. R., Lyttleton, O., McArdle, W. L., Ring, S. M., Horvath, S., Smith, G. D., & Relton, C. L. (2017). The epigenetic clock and physical development during childhood and adolescence: Longitudinal analysis from a UK birth cohort. *International Journal of Epidemiology, 46*(2), 549–558. https://doi.org/10.1093/ije/dyw307

Singer, W. (1999). Neuronal synchrony. *Neuron, 24*(1), 49–65. https://doi.org/10.1016/S0896-6273(00)80821-1

Singer, W. (2004). Verschaltungen legen uns fest: Wir sollten aufhören, von Freiheit zu sprechen. In C. Geyer (Ed.), *Hirnforschung Und Willensfreiheit* (pp. 30–65). Suhrkamp.

Singer, W. (2013). Cortical dynamics revisited. *Trends in Cognitive Sciences, 17*(12), 616–626. https://doi.org/10.1016/j.tics.2013.09.006

Singer, W. (2021). Recurrent dynamics in the cerebral cortex: Integration of sensory evidence with stored knowledge. *Proceedings of the National Academy of Sciences, 118*(33), e2101043118. https://doi.org/10.1073/pnas.2101043118

Singh, A. K., Phillips, F., Merabet, L. B., & Sinha, P. (2018). Why does the cortex reorganize after sensory loss? *Trends in Cognitive Sciences, 22*(7), 569–582. https://doi.org/10.1016/j.tics.2018.04.004

Sjöstedt, E., Zhong, W., Fagerberg, L., Karlsson, M., Mitsios, N., Adori, C., Oksvold, P., Edfors, F., Limiszewska, A., Hikmet, F., Huang, J., Du, Y., Lin, L., Dong, Z., Yang, L., Liu, X., Jiang, H., Xu, X., Wang, J., et al. (2020). An atlas of the protein-coding genes in the human, pig, and mouse brain. *Science, 367*(6482), eaay5947. https://doi.org/10.1126/science.aay5947

Södersten, E., Toskas, K., Rraklli, V., Tiklova, K., Björklund, Å. K., Ringnér, M., Perlmann, T., & Holmberg, J. (2018). A comprehensive map coupling histone modifications with gene regulation in adult dopaminergic and serotonergic neurons. *Nature Communications, 9*(1), Article 1. https://doi.org/10.1038/s41467-018-03538-9

Sohal, V. S., Zhang, F., Yizhar, O., & Deisseroth, K. (2009). Parvalbumin neurons and gamma rhythms enhance cortical circuit performance. *Nature, 459*(7247), 698–702. https://doi.org/10.1038/nature07991

Soler, L., Zwart, S. D., Israel-Jost, V., & Lynch, M. (Eds.). (2017). *Science after the practice turn in philosophy, history, and social studies of science* (First issued in paperback). Routledge, Taylor and Francis Group.

Solms, M. L. (2018). The neurobiological underpinnings of psychoanalytic theory and therapy. *Frontiers in Behavioral Neuroscience, 12*, 294. https://www.frontiersin.org/articles/10.3389/fnbeh.2018.00294

Soon, C., Brass, M., Heinze, H.-J., & Haynes, J.-D. (2008). Unconscious determinants of free decisions in the human brain. *Nature Neuroscience, 11*, 543–545. https://doi.org/10.1038/nn.2112

Sporns, O. (2013). Structure and function of complex brain networks. *Dialogues in Clinical Neuroscience, 15*(3), 247–262.

Sporns, O., Tononi, G., & Kötter, R. (2005). The human connectome: A structural description of the human brain. *PLoS Computational Biology, 1*(4), e42. https://doi.org/10.1371/journal.pcbi.0010042

Sroufe, L. A., Cooper, R. G., DeHart, G., & Bronfenbrenner, U. (1992). *Child development: Its nature and course* (2nd ed.). McGraw-Hill.

Stam, H. J. (2015). The neurosciences and the search for a unified psychology: The science and esthetics of a single framework. *Frontiers in Psychology, 6*, 1467. https://doi.org/10.3389/fpsyg.2015.01467

Stankiewicz, A. M., Swiergiel, A. H., & Lisowski, P. (2013). Epigenetics of stress adaptations in the brain. *Brain Research Bulletin, 98*, 76–92. https://doi.org/10.1016/j.brainresbull.2013.07.003

Stevens, B. (2003). Glia: Much more than the neuron's side-kick. *Current Biology, 13*(12), R469–R472. https://doi.org/10.1016/S0960-9822(03)00404-4

Stogsdill, J. A., & Eroglu, C. (2017). The interplay between neurons and glia in synapse development and plasticity. *Current Opinion in Neurobiology, 42*, 1–8. https://doi.org/10.1016/j.conb.2016.09.016

Stotz, K. (2017). Why developmental niche construction is not selective niche construction: And why it matters. *Interface Focus, 7*(5), 20160157. https://doi.org/10.1098/rsfs.2016.0157

Sui, J., Liu, C. H., & Han, S. (2009). Cultural difference in neural mechanisms of self-recognition. *Social Neuroscience, 4*(5), 402–411. https://doi.org/10.1080/17470910802674825

Sultan, F. A., & Day, J. J. (2011). Epigenetic mechanisms in memory and synaptic function. *Epigenomics, 3*(2), 157–181. https://doi.org/10.2217/epi.11.6

Sussillo, D. (2014). Neural circuits as computational dynamical systems. *Current Opinion in Neurobiology, 25*, 156–163. https://doi.org/10.1016/j.conb.2014.01.008

Sweatt, J. D. (2013). The emerging field of neuroepigenetics. *Neuron, 80*(3), 624–632. https://doi.org/10.1016/j.neuron.2013.10.023

Sweatt, J. D. (2016). Dynamic DNA methylation controls glutamate receptor trafficking and synaptic scaling. *Journal of Neurochemistry, 137*(3), 312–330. https://doi.org/10.1111/jnc.13564

Tal, Z., Geva, R., & Amedi, A. (2017). Positive and negative somatotopic BOLD responses in contralateral versus ipsilateral penfield homunculus. *Cerebral Cortex (New York, NY), 27*(2), 962–980. https://doi.org/10.1093/cercor/bhx024

Tang, Y., Zhang, W., Chen, K., Feng, S., Ji, Y., Shen, J., Reiman, E. M., & Liu, Y. (2006). Arithmetic processing in the brain shaped by cultures. *Proceedings of the National Academy of Sciences, 103*(28), 10775–10780. https://doi.org/10.1073/pnas.0604416103

Teles, R. V. (2020). Phineas Gage's great legacy. *Dementia & Neuropsychologia, 14*(4), 419–421. https://doi.org/10.1590/1980-57642020dn14-040013

ten Oever, S., & Sack, A. T. (2019). Interactions between rhythmic and feature predictions to create parallel time-content associations. *Frontiers in Neuroscience, 13*, 791. https://www.frontiersin.org/articles/10.3389/fnins.2019.00791

Teschendorff, A. E., & Zheng, S. C. (2017). Cell-type deconvolution in epigenome-wide association studies: A review and recommendations. *Epigenomics, 9*(5), 757–768. https://doi.org/10.2217/epi-2016-0153

Teschendorff, A. E., Yang, Z., Wong, A., Pipinikas, C. P., Jiao, Y., Jones, A., Anjum, S., Hardy, R., Salvesen, H. B., Thirlwell, C., Janes, S. M., Kuh, D., & Widschwendter, M. (2015). Correlation of smoking-associated DNA methylation changes in buccal cells with DNA methylation changes in epithelial cancer. *JAMA Oncology, 1*(4), 476–485. https://doi.org/10.1001/jamaoncol.2015.1053

Timme, N. M., & Lapish, C. (2018). A tutorial for information theory in neuroscience. *ENeuro, 5*(3), ENEURO.0052-18.2018. https://doi.org/10.1523/ENEURO.0052-18.2018

Todd, E. V., Liu, H., Muncaster, S., & Gemmell, N. J. (2016). Bending genders: The biology of natural sex change in fish. *Sexual Development, 10*(5–6), 223–241. https://doi.org/10.1159/000449297

Tognini, P., Napoli, D., & Pizzorusso, T. (2015). Dynamic DNA methylation in the brain: A new epigenetic mark for experience-dependent plasticity. *Frontiers in Cellular Neuroscience, 9*, 331. https://www.frontiersin.org/articles/10.3389/fncel.2015.00331

Tononi, G. (2004). An information integration theory of consciousness. *BMC Neuroscience, 5*(1), 42. https://doi.org/10.1186/1471-2202-5-42

Tononi, G., Boly, M., Massimini, M., & Koch, C. (2016). Integrated information theory: From consciousness to its physical substrate. *Nature Reviews Neuroscience, 17*(7), Article 7. https://doi.org/10.1038/nrn.2016.44

Truccolo, W., Eden, U. T., Fellows, M. R., Donoghue, J. P., & Brown, E. N. (2005). A point process framework for relating neural spiking activity to spiking history, neural ensemble, and extrinsic covariate effects. *Journal of Neurophysiology, 93*(2), 1074–1089. https://doi.org/10.1152/jn.00697.2004

Tsao, A., Sugar, J., Lu, L., Wang, C., Knierim, J. J., Moser, M.-B., & Moser, E. I. (2018). Integrating time from experience in the lateral entorhinal cortex. *Nature, 561*(7721), Article 7721. https://doi.org/10.1038/s41586-018-0459-6

Tuckute, G., Paunov, A., Kean, H., Small, H., Mineroff, Z., Blank, I., & Fedorenko, E. (2022). Frontal language areas do not emerge in the absence of temporal language areas: A case study of an individual born without a left temporal lobe. *Neuropsychologia, 169*, 108184. https://doi.org/10.1016/j.neuropsychologia.2022.108184

Turrigiano, G. G. (2017). The dialectic of Hebb and homeostasis. *Philosophical Transactions of the Royal Society of London. Series B, Biological Sciences, 372*(1715), 20160258. https://doi.org/10.1098/rstb.2016.0258

Turrigiano, G. G., & Nelson, S. B. (2000). Hebb and homeostasis in neuronal plasticity. *Current Opinion in Neurobiology, 10*(3), 358–364. https://doi.org/10.1016/s0959-4388(00)00091-x

Turrigiano, G. G., Leslie, K. R., Desai, N. S., Rutherford, L. C., & Nelson, S. B. (1998). Activity-dependent scaling of quantal amplitude in neocortical neurons. *Nature, 391*(6670), 892–896. https://doi.org/10.1038/36103

Uchida, S., Teubner, B. J. W., Hevi, C., Hara, K., Kobayashi, A., Dave, R. M., Shintaku, T., Jaikhan, P., Yamagata, H., Suzuki, T., Watanabe, Y., Zakharenko, S. S., & Shumyatsky, G. P. (2017). CRTC1 nuclear translocation following learning modulates memory strength via exchange of chromatin remodeling complexes on the Fgf1 gene. *Cell Reports, 18*(2), 352–366. https://doi.org/10.1016/j.celrep.2016.12.052

Uhlén, M., Fagerberg, L., Hallström, B. M., Lindskog, C., Oksvold, P., Mardinoglu, A., Sivertsson, Å., Kampf, C., Sjöstedt, E., Asplund, A., Olsson, I., Edlund, K., Lundberg, E., Navani, S., Szigyarto, C. A.-K., Odeberg, J., Djureinovic, D., Takanen, J. O., Hober, S., et al. (2015). Tissue-based map of the human proteome. *Science, 347*(6220), 1260419. https://doi.org/10.1126/science.1260419

Uhlhaas, P., Pipa, G., Lima, B., Melloni, L., Neuenschwander, S., Nikolić, D., & Singer, W. (2009). Neural synchrony in cortical networks: History, concept and current status. *Frontiers in Integrative Neuroscience, 3*, 17. https://www.frontiersin.org/articles/10.3389/neuro.07.017.2009

Urendes Jiménez, E., Flores Caballero, A., Molina Rueda, F., Iglesias Giménez, J., & Oboe, R. (2014). Reverse-engineer the brain: Perspectives and challenges. In J. L. Pons & D. Torricelli (Eds.), *Emerging therapies in neurorehabilitation* (pp. 173–188). Springer. https://doi.org/10.1007/978-3-642-38556-8_9

Vaaga, C. E., Borisovska, M., & Westbrook, G. L. (2014). Dual-transmitter neurons: Functional implications of co-release and co-transmission. *Current Opinion in Neurobiology, 0*, 25–32. https://doi.org/10.1016/j.conb.2014.04.010

Valsiner, J. (2007). Gilbert Gottlieb's theory of probabilistic epigenesis: Probabilities and realities in development. *Developmental Psychobiology, 49*(8), 832–840. https://doi.org/10.1002/dev.20276

Valsiner, J. (2014). Needed for cultural psychology: Methodology in a new key. *Culture & Psychology, 20*(1), 3–30. https://doi.org/10.1177/1354067X13515941

Valsiner, J. (2022a). Pathways to generalization: General knowledge as abstract complementation. In J. Valsiner (Ed.), *One dog is enough: Ivan P. Pavlov's contribution to ideographic science* (pp. 47–68). Information Age Publishing.

Valsiner, J. (2022b). Return to the dog: Ivan P. Pavlov as a pioneer of idiographic science. In J. Valsiner (Ed.), *One dog is enough: Ivan P. Pavlov's contribution to ideographic science* (pp. vii–xiv). Information Age Publishing.

Van Gebuchten, A. (1891). La structure des centres nerveux. La moiille epiniere et le cervelet. *Cellule, 7*, 79–122.

van Hemmen, J. L., & Sejnowski, T. J. (Eds.). (2006). *23 Problems in systems neuroscience* (pp. xvi, 514). Oxford University Press. https://doi.org/10.1093/acprof:oso/9780195148220.001.0001

VanLeeuwen, J.-E., Rafalovich, I., Sellers, K., Jones, K. A., Griffith, T. N., Huda, R., Miller, R. J., Srivastava, D. P., & Penzes, P. (2014). Coordinated nuclear and synaptic shuttling of afadin promotes spine plasticity and histone modifications. *The Journal of Biological Chemistry, 289*(15), 10831–10842. https://doi.org/10.1074/jbc.M113.536391

Varela, F. J., Thompson, E., & Rosch, E. (2016). *The embodied mind: Cognitive science and human experience* (Revised edition). MIT Press.

Vieira, M. S., Goulart, V. A. M., Parreira, R. C., Oliveira-Lima, O. C., Glaser, T., Naaldijk, Y. M., Ferrer, A., Savanur, V. H., Reyes, P. A., Sandiford, O., Rameshwar, P., Ulrich, H., Pinto, M. C. X., & Resende, R. R. (2019). Decoding epigenetic cell signaling in neuronal differentiation. *Seminars in Cell & Developmental Biology, 95*, 12–24. https://doi.org/10.1016/j.semcdb.2018.12.006

Viitaniemi, H. M., Verhagen, I., Visser, M. E., Honkela, A., van Oers, K., & Husby, A. (2019). Seasonal variation in genome-wide DNA methylation patterns and the onset of seasonal timing of reproduction in great tits. *Genome Biology and Evolution, 11*(3), 970–983. https://doi.org/10.1093/gbe/evz044

Vöhringer, M. (2009). Reflex. Begriff und Experiment. In *Begriffsgeschichte der Naturwissenschaften* (pp. 203–214). De Gruyter. https://doi.org/10.1515/9783110213034.3.203

von der Malsburg, C. (1994). The correlation theory of brain function. In E. Domany, J. L. van Hemmen, & K. Schulten (Eds.), *Models of neural networks: Temporal aspects of coding and information processing in biological systems* (pp. 95–119). Springer. https://doi.org/10.1007/978-1-4612-4320-5_2

von Kölliker, A. (1890). *Zur feineren Anatomie des centralen Nervensystems*. Wilhelm Engelmann. https://wellcomecollection.org/works/ycqnc9cv

von Uexküll, J. (1926). *Theoretical biology*. K. Paul, Trench, Trubner & Co. Ltd.

von Uexküll, J. (2010). *A foray into the worlds of animals and humans: With A theory of meaning* (1st University of Minnesota Press ed.). University of Minnesota Press. http://site.ebrary.com/id/10442224

von Uexküll, T. (1992). Introduction: The sign theory of Jakob von Uexküll. *Semiotica, 89*(4), 279–316. https://doi.org/10.1515/semi.1992.89.4.279

Vujovic, V., Rosendo, A., Brodbeck, L., & Iida, F. (2017). Evolutionary developmental robotics: Improving morphology and control of physical robots. *Artificial Life, 23*(2), 169–185. https://doi.org/10.1162/ARTL_a_00228

Wada, K., Hayase, S., Imai, R., Mori, C., Kobayashi, M., Liu, W., Takahasi, M., & Okanoya, K. (2013). Differential androgen receptor expression and DNA methylation state in striatum song nucleus Area X between wild and domesticated songbird strains. *European Journal of Neuroscience, 38*(4), 2600–2610. https://doi.org/10.1111/ejn.12258

Waddington, C. H. (1940). *Organisers & genes*. The University Press.

Waddington, C. H. (1942). The epigenotype. *Endeavour, 1*, 18–20.

Waddington, C. H. (1953). Genetic assimilation of an acquired character. *Evolution, 7*(2), 118–126. https://doi.org/10.2307/2405747

Waddington, C. H. (1957). *The strategy of the genes: A discussion of some aspects of theoretical biology: With an appendix by H. Kacser*. George Allen and Unwin.

Wadhwa, P. D., Buss, C., Entringer, S., & Swanson, J. M. (2009). Developmental origins of health and disease: Brief history of the approach and current focus on epigenetic mechanisms. *Seminars in Reproductive Medicine, 27*(5), 358–368. https://doi.org/10.1055/s-0029-1237424

Wagner, A. (1999). Causality in complex systems. *Biology and Philosophy, 14*(1), 83–101. https://doi.org/10.1023/A:1006580900476

Wagner, H., & Gaese, B. (2006). Can we understand the action of brains in natural environments? In J. L. van Hemmen & T. J. Sejnowski (Eds.), *23 Problems in systems neuroscience* (pp. 22–43). Oxford University Press. https://doi.org/10.1093/acprof:oso/9780195148220.003.0002

Waldeyer, W. (1891). *Ueber einige neuere Forschungen im Gebiete der Anatomie des Centralnervensystems*. Verlag von Georg Thieme.

Walter, J., & Schickl, H. (Eds.). (2019). *Single-cell analysis in research and medicine. Report of the Interdisciplinary Research Group Gene Technology Report*. Berlin-Brandenburg Academy of Sciences and Humanities.

Wang, B., & Dudko, O. K. (2021). A theory of synaptic transmission. *eLife, 10*, e73585. https://doi.org/10.7554/eLife.73585

Wang, X., & Moazed, D. (2017). DNA sequence-dependent epigenetic inheritance of gene silencing and histone H3K9 methylation. *Science, 356*(6333), 88–91. https://doi.org/10.1126/science.aaj2114

Wang, Z., Tang, B., He, Y., & Jin, P. (2016). DNA methylation dynamics in neurogenesis. *Epigenomics, 8*(3), 401–414. https://doi.org/10.2217/epi.15.119

Wang, T., Tsui, B., Kreisberg, J. F., Robertson, N. A., Gross, A. M., Yu, M. K., Carter, H., Brown-Borg, H. M., Adams, P. D., & Ideker, T. (2017). Epigenetic aging signatures in mice livers are slowed by dwarfism, calorie restriction and rapamycin treatment. *Genome Biology, 18*(1), 57. https://doi.org/10.1186/s13059-017-1186-2

Wang, T., Morency, D. T., Harris, N., & Davis, G. W. (2020). Epigenetic signaling in glia controls presynaptic homeostatic plasticity. *Neuron, 105*(3), 491–505.e3. https://doi.org/10.1016/j.neuron.2019.10.041

Watanabe, S. (2015). Slow or fast? A tale of synaptic vesicle recycling. *Science, 350*(6256), 46–47. https://doi.org/10.1126/science.aad2996

Waterland, R. A., & Jirtle, R. L. (2003). Transposable elements. *Molecular and Cellular Biology, 23*(15), 5293–5300. https://doi.org/10.1128/MCB.23.15.5293-5300.2003

Weaver, I. C. G., Cervoni, N., Champagne, F. A., D'Alessio, A. C., Sharma, S., Seckl, J. R., Dymov, S., Szyf, M., & Meaney, M. J. (2004). Epigenetic programming by maternal behavior. *Nature Neuroscience, 7*(8), 847–854. https://doi.org/10.1038/nn1276

Webster, A. P., Plant, D., Ecker, S., Zufferey, F., Bell, J. T., Feber, A., Paul, D. S., Beck, S., Barton, A., Williams, F. M. K., & Worthington, J. (2018). Increased DNA methylation variability in rheumatoid arthritis-discordant monozygotic twins. *Genome Medicine, 10*(1), 64. https://doi.org/10.1186/s13073-018-0575-9

Wei, J., Xiong, Z., Lee, J. B., Cheng, J., Duffney, L. J., Matas, E., & Yan, Z. (2016). Histone modification of Nedd4 ubiquitin ligase controls the loss of AMPA receptors and cognitive impairment induced by repeated stress. *The Journal of Neuroscience: The Official Journal of the Society for Neuroscience, 36*(7), 2119–2130. https://doi.org/10.1523/JNEUROSCI.3056-15.2016

Weidner, C. I., Lin, Q., Koch, C. M., Eisele, L., Beier, F., Ziegler, P., Bauerschlag, D. O., Jöckel, K.-H., Erbel, R., Mühleisen, T. W., Zenke, M., Brümmendorf, T. H., & Wagner, W. (2014). Aging of blood can be tracked by DNA methylation changes at just three CpG sites. *Genome Biology, 15*(2), R24. https://doi.org/10.1186/gb-2014-15-2-r24

Weigel, S. (2016). Embodiment in simulation theory and cultural science, with remarks on the coding-problem of neuroscience. In S. Weigel & G. Scharbert (Eds.), *A neuro-psychoanalytical dialogue for bridging Freud and the neurosciences* (pp. 47–71). Springer International Publishing. https://doi.org/10.1007/978-3-319-17605-5_4

Wendelken, C., Ferrer, E., Whitaker, K. J., & Bunge, S. A. (2016). Fronto-parietal network reconfiguration supports the development of reasoning ability. *Cerebral Cortex (New York, N.Y.: 1991), 26*(5), 2178–2190. https://doi.org/10.1093/cercor/bhv050

West-Eberhard, M. J. (2003). *Developmental plasticity and evolution*. Oxford University Press.

West-Eberhard, M. J. (2005). Developmental plasticity and the origin of species differences. *Proceedings of the National Academy of Sciences, 102*(suppl_1), 6543–6549. https://doi.org/10.1073/pnas.0501844102

Wilson, M. (2002). Six views of embodied cognition. *Psychonomic Bulletin & Review, 9*(4), 625–636. https://doi.org/10.3758/BF03196322

Winograd, S., & Cowan, J. D. (1963). *Reliable computation in the presence of noise* (X956.88). MIT Press; Computer History Museum.

Woltereck, R. (1913). Weitere experimentelle untersuchungen über Artänderung, speziell über das Wesen quantitativer Artunterschiede bei Daphniden. *Zeitschrift für Induktive Abstammungs- und Vererbungslehre, 9*(1), 146–146. https://doi.org/10.1007/BF01876686

Wong, V. S., & Langley, B. (2016). Epigenetic changes following traumatic brain injury and their implications for outcome, recovery and therapy. *Neuroscience Letters, 625*, 26–33. https://doi.org/10.1016/j.neulet.2016.04.009

Wright, B. E. (1979). Causality in biological systems. *Trends in Biochemical Sciences, 4*(5), N110–N111. https://doi.org/10.1016/0968-0004(79)90388-8

Wundt, W. (1888). Selbstbeobachtung und innere Wahrnehmung. *Philosophische Studien, 4*, 292–309.

Wundt, W. (1897). *Outline of psychology* (pp. xviii, 342). Wilhelm Engelmann. https://doi.org/10.1037/12908-000

Wundt, W. (1904). *Principles of physiological psychology*. Swan Sonnenschein & Co. Lim.

Wytock, T. P., & Motter, A. E. (2020). Distinguishing cell phenotype using cell epigenotype. *Science Advances, 6*(12), eaax7798. https://doi.org/10.1126/sciadv.aax7798

Xia, L., Ma, S., Zhang, Y., Wang, T., Zhou, M., Wang, Z., & Zhang, J. (2015). Daily variation in global and local DNA methylation in mouse livers. *PLoS One, 10*(2), e0118101. https://doi.org/10.1371/journal.pone.0118101

Xiong, H., & Gendelman, H. E. (Eds.). (2014). *Current laboratory methods in neuroscience research*. Springer.

Xu, Y., Jia, Y., Ma, J., Hayat, T., & Alsaedi, A. (2018). Collective responses in electrical activities of neurons under field coupling. *Scientific Reports, 8*(1), Article 1. https://doi.org/10.1038/s41598-018-19858-1

Yamada, L., & Chong, S. (2017). Epigenetic studies in developmental origins of health and disease: Pitfalls and key considerations for study design and interpretation. *Journal of Developmental Origins of Health and Disease, 8*(1), 30–43. https://doi.org/10.1017/S2040174416000507

Yamamuro, K., Kimoto, S., Rosen, K. M., Kishimoto, T., & Makinodan, M. (2015). Potential primary roles of glial cells in the mechanisms of psychiatric disorders. *Frontiers in Cellular Neuroscience, 9*, 154. https://doi.org/10.3389/fncel.2015.00154

Yang, Z., Wong, A., Kuh, D., Paul, D. S., Rakyan, V. K., Leslie, R. D., Zheng, S. C., Widschwendter, M., Beck, S., & Teschendorff, A. E. (2016). Correlation of an epigenetic mitotic clock with cancer risk. *Genome Biology, 17*(1), 205. https://doi.org/10.1186/s13059-016-1064-3

Yehuda, R., & Lehrner, A. (2018). Intergenerational transmission of trauma effects: Putative role of epigenetic mechanisms. *World Psychiatry, 17*(3), 243–257. https://doi.org/10.1002/wps.20568

Yehuda, R., Daskalakis, N. P., Bierer, L. M., Bader, H. N., Klengel, T., Holsboer, F., & Binder, E. B. (2016). Holocaust exposure induced intergenerational effects on FKBP5 methylation. *Biological Psychiatry, 80*(5), 372–380. https://doi.org/10.1016/j.biopsych.2015.08.005

Yehuda, R., Lehrner, A., & Bierer, L. M. (2018). The public reception of putative epigenetic mechanisms in the transgenerational effects of trauma. *Environmental Epigenetics, 4*(2), dvy018. https://doi.org/10.1093/eep/dvy018

Yin, W., Li, T., Hung, S.-C., Zhang, H., Wang, L., Shen, D., Zhu, H., Mucha, P. J., Cohen, J. R., & Lin, W. (2020). The emergence of a functionally flexible brain during early infancy. *Proceedings of the National Academy of Sciences, 117*(38), 23904–23913. https://doi.org/10.1073/pnas.2002645117

Yu, F., Jiang, Q., Sun, X., & Zhang, R. (2015). A new case of complete primary cerebellar agenesis: Clinical and imaging findings in a living patient. *Brain, 138*(6), e353. https://doi.org/10.1093/brain/awu239

Yuste, R., Hawrylycz, M., Aalling, N., Aguilar-Valles, A., Arendt, D., Armañanzas, R., Ascoli, G. A., Bielza, C., Bokharaie, V., Bergmann, T. B., Bystron, I., Capogna, M., Chang, Y., Clemens, A., de Kock, C. P. J., DeFelipe, J., Dos Santos, S. E., Dunville, K., Feldmeyer, D., et al. (2020). A community-based transcriptomics classification and nomenclature of neocortical cell types. *Nature Neuroscience, 23*(12), Article 12. https://doi.org/10.1038/s41593-020-0685-8

Zarrinkoob, L., Ambarki, K., Wåhlin, A., Birgander, R., Eklund, A., & Malm, J. (2015). Blood flow distribution in cerebral arteries. *Journal of Cerebral Blood Flow & Metabolism, 35*(4), 648–654. https://doi.org/10.1038/jcbfm.2014.241

Zbieć-Piekarska, R., Spólnicka, M., Kupiec, T., Parys-Proszek, A., Makowska, Ż., Pałeczka, A., Kucharczyk, K., Płoski, R., & Branicki, W. (2015). Development of a forensically useful age prediction method based on DNA methylation analysis. *Forensic Science International. Genetics, 17*, 173–179. https://doi.org/10.1016/j.fsigen.2015.05.001

Zeharia, N., Hertz, U., Flash, T., & Amedi, A. (2012). Negative blood oxygenation level dependent homunculus and somatotopic information in primary motor cortex and supplementary motor area. *Proceedings of the National Academy of Sciences of the United States of America, 109*, 18565–18570. https://doi.org/10.1073/pnas.1119125109

Zeharia, N., Hertz, U., Flash, T., & Amedi, A. (2015). New whole-body sensory-motor gradients revealed using phase-locked analysis and verified using multivoxel pattern analysis and functional connectivity. *Journal of Neuroscience, 35*(7), 2845–2859. https://doi.org/10.1523/JNEUROSCI.4246-14.2015

Zhang, X., & Ho, S.-M. (2011). Epigenetics meets endocrinology. *Journal of Molecular Endocrinology, 46*(1), R11–R32.

Zhang, W., & Linden, D. J. (2003). The other side of the engram: Experience-driven changes in neuronal intrinsic excitability. *Nature Reviews Neuroscience, 4*(11), Article 11. https://doi.org/10.1038/nrn1248

Zhang, Q., Vallerga, C. L., Walker, R. M., Lin, T., Henders, A. K., Montgomery, G. W., He, J., Fan, D., Fowdar, J., Kennedy, M., Pitcher, T., Pearson, J., Halliday, G., Kwok, J. B., Hickie, I., Lewis, S., Anderson, T., Silburn, P. A., Mellick, G. D., et al. (2019). Improved precision of epigenetic clock estimates across tissues and its implication for biological ageing. *Genome Medicine, 11*(1), 54. https://doi.org/10.1186/s13073-019-0667-1

Zhao, T. C., Llanos, F., Chandrasekaran, B., & Kuhl, P. K. (2022). Language experience during the sensitive period narrows infants' sensory encoding of lexical tones-Music intervention reverses it. *Frontiers in Human Neuroscience, 16*, 941853. https://doi.org/10.3389/fnhum.2022.941853

Zhu, T., Liu, J., Beck, S., Pan, S., Capper, D., Lechner, M., Thirlwell, C., Breeze, C. E., & Teschendorff, A. E. (2022). A pan-tissue DNA methylation atlas enables in silico decomposition of human tissue methylomes at cell-type resolution. *Nature Methods, 19*(3), Article 3. https://doi.org/10.1038/s41592-022-01412-7

Ziemann, U., Ilić, T. V., & Jung, P. (2006). Chapter 3: Long-term potentiation (LTP)-like plasticity and learning in human motor cortex – Investigations with transcranial magnetic stimulation (TMS). In C. Barber, S. Tsuji, S. Tobimatsu, T. Uozumi, N. Akamatsu, & A. Eisen (Eds.), *Supplements to clinical neurophysiology* (Vol. 59, pp. 19–25). Elsevier. https://doi.org/10.1016/S1567-424X(09)70007-8

Index